# GCSE separate sciences

**Authors**

Michael Brimicombe

Simon Broadley

Philippa Gardom-Hulme

Mark Matthews

# How to use this book

Welcome to your AQA GCSE separate sciences revision guide. This book has been specially written by experienced teachers and examiners to match the 2011 specification.

On this page you can see the types of feature you will find in this book. Everything in the book is designed to provide you with the support you need to help you prepare for your examinations and achieve your best.

**Specification and student book reference**: These show how the pages in the unit match to the exam specification and to your textbook so you can track your progress through the unit as you learn.

**Key words**: These are the terms you need to understand for your exams.

**Exam tip**: These hints will help you to think about what may come up in the exam.

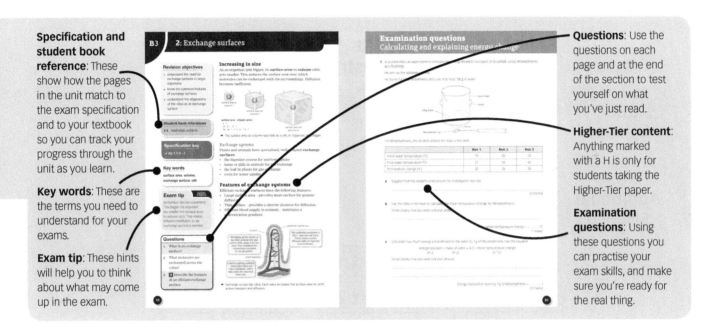

**Questions**: Use the questions on each page and at the end of the section to test yourself on what you've just read.

**Higher-Tier content**: Anything marked with a H is only for students taking the Higher-Tier paper.

**Examination questions**: Using these questions you can practise your exam skills, and make sure you're ready for the real thing.

**Upgrade**: Upgrade takes you through an exam question in a step-by-step way, showing you why different answers get different grades. Using the tips on the page you can make sure you achieve your best by understanding what each question needs and what an examiner is looking for in your answer.

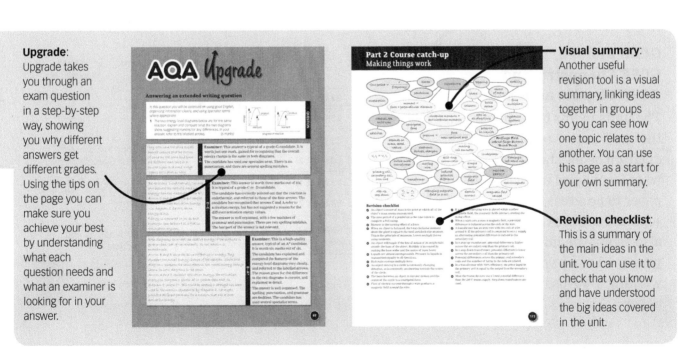

**Visual summary**: Another useful revision tool is a visual summary, linking ideas together in groups so you can see how one topic relates to another. You can use this page as a start for your own summary.

**Revision checklist**: This is a summary of the main ideas in the unit. You can use it to check that you know and have understood the big ideas covered in the unit.

## Matching your course

The units in this book have been written to match the specifications, so that you can take three separate GCSEs in science.

In the diagram below you can see that the units and part units can be used to study either for **GCSE Science**, leading to GCSE Additional Science, or as part of GCSE Biology, GCSE Chemistry, and GCSE Physics courses.

|  | GCSE Biology | GCSE Chemistry | GCSE Physics |
|---|---|---|---|
| **GCSE Science** | B1 (Part 1) | C1 (Part 1) | P1 (Part 1) |
|  | B1 (Part 2) | C1 (Part 2) | P1 (Part 2) |
| **GCSE Additional Science** | B2 (Part 1) | C2 (Part 1) | P2 (Part 1) |
|  | B2 (Part 2) | C2 (Part 2) | P2 (Part 2) |
|  | **B3 (Part 1)** | **C3 (Part 1)** | **P3 (Part 1)** |
|  | **B3 (Part 2)** | **C3 (Part 2)** | **P3 (Part 2)** |

**GCSE Biology, GCSE Chemistry, and GCSE Physics assessment**

The units in this book match the exam papers on offer. The diagram below shows you what is included in each exam paper to progress to three separate science GCSEs. It also shows you how much of your final mark you will be working towards in each paper.

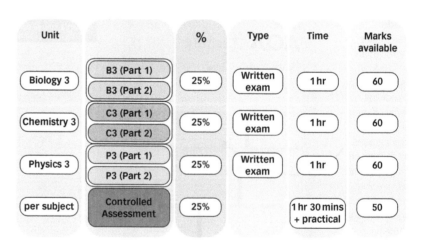

| Unit | | % | Type | Time | Marks available |
|---|---|---|---|---|---|
| Biology 3 | B3 (Part 1) / B3 (Part 2) | 25% | Written exam | 1 hr | 60 |
| Chemistry 3 | C3 (Part 1) / C3 (Part 2) | 25% | Written exam | 1 hr | 60 |
| Physics 3 | P3 (Part 1) / P3 (Part 2) | 25% | Written exam | 1 hr | 60 |
| per subject | Controlled Assessment | 25% | | 1 hr 30 mins + practical | 50 |

When you read the questions in your exam papers you should make sure you know what kind of answer you are being asked for. The list below explains some of the common words you will see used in exam questions. Make sure you know what each word means. Always read the question thoroughly, even if you recognise the word used.

## Calculate

Work out your answer by using a calculation. You can use your calculator to help you. You may need to use an equation; check whether one has been provided for you in the paper. The question will say if your working must be shown.

## Describe

Write a detailed answer that covers what happens, when it happens, and where it happens. The question will let you know how much of the topic to cover. Talk about facts and characteristics. (Hint: don't confuse with 'Explain')

## Explain

You will be asked how or why something happens. Write a detailed answer that covers how and why a thing happens. Talk about mechanisms and reasons. (Hint: don't confuse with 'Describe')

## Evaluate

You will be given some facts, data or other information. Write about the data or facts and provide your own conclusion or opinion on them.

## Outline

Give only the key facts of the topic. You may need to set out the steps of a procedure or process – make sure you write down the steps in the correct order.

## Show

Write down the details, steps or calculations needed to prove an answer that you have been given.

## Suggest

Think about what you've learnt in your science lessons and apply it to a new situation or a context. You may not know the answer. Use what you have learnt to suggest sensible answers to the question.

## Write down

Give a short answer, without a supporting argument.

### Top tips

Always read exam questions carefully, even if you recognise the word used. Look at the information in the question and the number of answer lines to see how much detail the examiner is looking for.

You can use bullet points or a diagram if it helps your answer.

If a number needs units you should include them, unless the units are already given on the answer line.

## Revision objectives

- ✓ understand the process of osmosis
- ✓ know how exercise and sports drinks affect the hydration of the body
- ✓ know how the process of active transport occurs

## Specification key

✓ B3.1.1 a – g

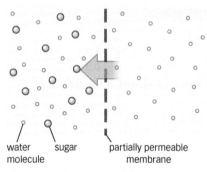

water molecule    sugar    partially permeable membrane

▲ Water is separated from a sugar solution by a partially permeable membrane. Sugar is too big to fit through the membrane pores but the water will pass through. So water molecules move into the sugar solution to dilute it.

# Movement of molecules

Molecules can move into and out of cells by three methods. Dissolved substances move by:

- diffusion
- active transport.

Water moves by:

- **osmosis**.

# Osmosis

Osmosis is a special kind of diffusion, which involves the movement of water only, into or out of cells.

During osmosis:

- water moves
- from an area of high water concentration (a dilute solution)
- to an area of low water concentration (a concentrated solution)
- through a **partially permeable membrane**
- until the concentrations even out.

## Osmosis and cells

The movement of water into and out of cells by osmosis is important for both plant and animal cells because it keeps their water levels in balance.

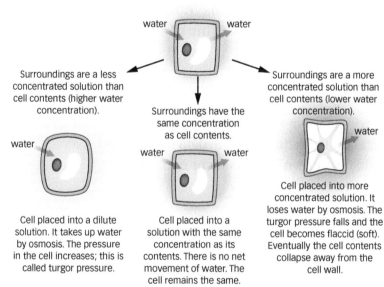

Surroundings are a less concentrated solution than cell contents (higher water concentration).

Surroundings have the same concentration as cell contents.

Surroundings are a more concentrated solution than cell contents (lower water concentration).

Cell placed into a dilute solution. It takes up water by osmosis. The pressure in the cell increases; this is called turgor pressure.

Cell placed into a solution with the same concentration as its contents. There is no net movement of water. The cell remains the same.

Cell placed into more concentrated solution. It loses water by osmosis. The turgor pressure falls and the cell becomes flaccid (soft). Eventually the cell contents collapse away from the cell wall.

▲ Water movement by osmosis in plant cells.

# The body and water

The human body contains a lot of water, which moves into the cells by osmosis. If the body's water level falls, the cells become **dehydrated**. It is important to replace lost water in the body by drinking.

A lot of water is lost from the body by sweating. This helps to cool us down. When we sweat we lose not only water but also ions from our bodies. The body's cells need water and ions to function correctly. Water and ions are also used in the body to:

- lubricate joints
- protect organs like the brain
- carry substances around the body
- help regulate body temperature.

## Sports drinks

During exercise we sweat more to cool our body; this results in a greater loss of water and ions. In addition we also use a lot of energy, which we release from glucose.

Sports drinks are designed to solve these problems. However, the drinks vary and should be used for different situations.

## Active transport

Some dissolved molecules or ions need to move into or out of cells from a low concentration to a high concentration, against a **concentration gradient**. This happens by a process called **active transport**. Active transport allows cells to absorb ions from very dilute solutions. Examples of active transport are:

- the uptake of ions by root hairs
- the movement of sodium ions out of nerve cells.

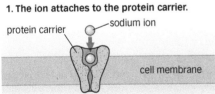

**1. The ion attaches to the protein carrier.**

protein carrier — sodium ion

cell membrane

There is a high concentration of sodium ions on the outside of the nerve, and a low concentration on the inside.

There are **protein carriers** in the nerve cell membrane. Sodium ions fit into these carriers.

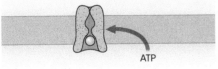

**2. The protein uses energy to change shape.**

ATP

The proteins can change shape. This uses energy from a molecule called ATP, which is made in respiration.

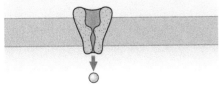

**3. The ion moves to the outside of the cell.**

As the protein changes shape, it moves the sodium ion from the inside of the membrane to the outside.

The sodium ion falls off the protein carrier.

The protein immediately returns to its normal shape.

## Key words

osmosis, partially permeable membrane, dehydrated, concentration gradient, active transport, protein carrier

**Water** to hydrate the body. Where dehydration is a problem, the more dilute drinks will hydrate the body quicker.

**Caffeine** to make us more alert.

**Ions** are dissolved minerals to keep the muscles healthy. Ions are lost during sweating. Sports drinks should contain ions at the correct body levels, so the body will only absorb ions up to that level.

**Carbohydrate** for energy. Moderate levels should be present to replace the glucose used during respiration. However, some drinks, called power drinks, have too much sugar and cause a sugar rush.

ENERGY

## Exam tip  AQA

Learn the definition of osmosis. There are four key phrases in the definition, and many questions can be answered by using these four points.

## Questions

1. What moves into a cell by osmosis?

2. What do sports drinks contain?

3. **H** What are the differences between osmosis and active transport?

## Working to Grade E

1. When we exercise will the following increase, decrease, or stay the same?
   a   sweating
   b   temperature
   c   water intake
2. Name **three** ways molecules get into cells.
3. On a hot sunny day are you more likely to:
   a   sweat **more/less**
   b   become **hydrated/dehydrated**
   c   become **more/less** thirsty
4. Does a dilute solution have a high or low water concentration?

## Working to Grade C

5. Define osmosis.
6. What is a partially permeable membrane?
7. Look at the diagram below.
   a   Indicate which direction the water will move.
   b   Explain why the movement will occur in this direction.

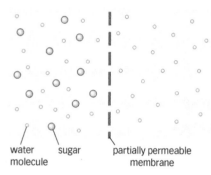

water        sugar          partially permeable
molecule                    membrane

8. Give **three** reasons why it is important to keep the body hydrated.
9. Apart from water, what else does the body lose during sweating?
10. Sports drinks contain a number of different ingredients. Complete the following table to explain the function of each ingredient.

| Ingredient | Function of ingredient |
|---|---|
| water | |
| sugar (carbohydrate) | |
| ions | |
| caffeine | |

11. Suggest an advantage of drinking sports drinks rather than water during exercise.

## Working to Grade A*

12. Look at the diagram in question 7. Explain why the sugar molecule does not move.
13. Describe what will happen to a potato chip after being placed for 24 hours in:
   a   distilled water
   b   a solution the same concentration as the cells
   c   a very strong sugar solution.
14. Below is a table of the ingredients found in three different sports drinks. Look at the data and answer the questions that follow.

| Drink | Sports Lite | Sports Active | Sports Power |
|---|---|---|---|
| water | 250 cm³ | 250 cm³ | 250 cm³ |
| glucose | 2.5 g | 8.7 g | 22.0 g |
| ions | 0.5 g | 0.5 g | 0. g |
| caffeine | 0.0 g | 0.0 g | 3.0 g |

   a   Which of the three drinks would be better for hydration?
   b   Explain why.
   c   Explain why Sports Power would be better for a sprinter just before a race.
   d   Explain why many people regard sports power drinks as unwise to take before sport.
15. Active transport is another form of transport.
   a   Complete the paragraph below by using the most suitable words to fill in the blanks:
       Some dissolved molecules or ions move from a _____ concentration to a _____ concentration. The movement is _____ a _____ gradient. This process is called active transport.
   b   Give **two** examples where active transport is used in living things.
   c   For each example state clearly:
       i     what is transported
       ii    from where
       iii   to where.
16. What **two** things are required for active transport to take place?
17. Explain why active transport will stop if a cell dies, or is given a respiratory poison.

# Examination questions
## Osmosis, sports drinks, and active transport

**1** A student set up an experiment where they placed a sugar solution inside a visking tubing bag and securely knotted the end. They then weighed the bag. The bag was placed into a beaker of distilled water for 30 minutes.

visking tubing bag
pure water
10% sugar solution

**a** Describe what the result of the experiment would be after 30 minutes.

.............................................................................................

.............................................................................................

*(1 mark)*

**b** Explain the reason for this result.

.......................................................................................................................

.......................................................................................................................

.......................................................................................................................

.......................................................................................................................

.......................................................................................................................

*(3 marks)*

*(Total marks: 4)*

**2** Sports drinks are used by athletes.

**a** Complete the following paragraph using some of the words supplied.

| glucose muscles energy fats hydrate dehydrate |

Sports drinks contain carbohydrates, water, and ions. Carbohydrates such as ................. are used in

respiration to provide ................. Water is needed to ................. the body. Mineral ions keep the

................. healthy.

*(4 marks)*

**b** The table shows some of the contents of two different sports drinks. Indicate by using a tick (✓) to show what these sports drinks are best used for.

| Drink contents | Hydration of the body | Energy supply for the body |
|---|---|---|
| Low sugar Dilute drink | | |
| High sugar Concentrated drink | | |

*(1 mark)*

**c** How is water lost from the body during exercise?

.......................................................................................................................

*(1 mark)*

*(Total marks: 6)*

# Revision objectives

- ✔ understand the need for exchange surfaces in larger organisms
- ✔ know the common features of exchange surfaces
- ✔ understand the adaptations of the villus as an exchange surface

## Student book references

3.4   Exchange surfaces

## Specification key

✔ B3.1.1 h – l

## Key words

surface area, volume, exchange surface, villi

## Exam tip

Remember this key statement: 'The bigger the organism, the smaller the surface-area-to-volume ratio.' This makes diffusion inefficient, so an exchange surface is needed.

## Questions

1   What is an exchange surface?

2   What molecules are exchanged across the villus?

3   H Describe the features of an efficient exchange surface.

# Increasing in size

As an organism gets bigger, its **surface-area**-to-**volume** ratio gets smaller. This reduces the surface area over which molecules can be exchanged with the surroundings. Diffusion becomes inefficient.

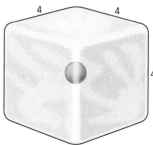

surface area 6
volume 1

surface area 24
volume 8

surface area 96
volume 64

surface area : volume ratios
  6 : 1
  24 : 8 = 3 : 1
  96 : 64 = 3 : 2 or 1.5 : 1

▲ The surface-area-to-volume ratio falls as a cell, or organism, gets bigger.

## Exchange systems

Plants and animals have specialised, well-adapted **exchange surfaces**:

- the digestive system for nutrient uptake
- lungs or gills in animals for gas exchange
- the leaf in plants for gas exchange
- roots for water uptake.

# Features of exchange systems

Efficient exchange surfaces have the following features:

- Large surface area – provides more surface for greater diffusion.
- Thin surface – provides a shorter distance for diffusion.
- Efficient blood supply in animals – maintains a concentration gradient.

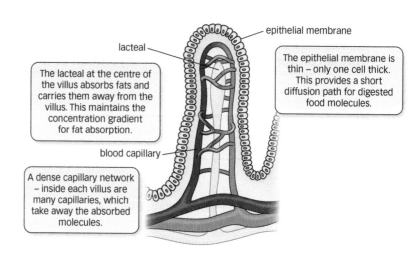

epithelial membrane

lacteal

The lacteal at the centre of the villus absorbs fats and carries them away from the villus. This maintains the concentration gradient for fat absorption.

The epithelial membrane is thin – only one cell thick. This provides a short diffusion path for digested food molecules.

blood capillary

A dense capillary network – inside each villus are many capillaries, which take away the absorbed molecules.

▲ Exchange across the villus. Each villus increases the surface area for both active transport and diffusion.

## The lungs

The **lungs** are located in the chest or **thorax**, surrounded by the rib cage. The ribs protect the lungs and are also used in the process of breathing. The thorax is separated from the **abdomen** by a muscular sheet called the **diaphragm**. This encloses the lungs in the thorax.

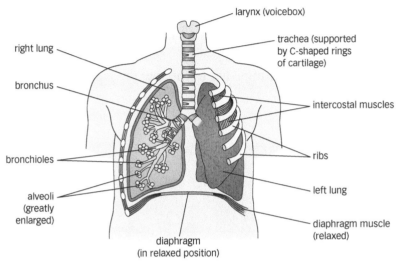

▲ This diagram shows a surface view of the left lung and a section through the right lung showing the airways and air sacs inside.

### Into the lungs

Air gets into the lungs.

- Air moves in through the nose and mouth.
- It passes into a tube called the windpipe or trachea.
- The trachea branches into two tubes, each called a bronchus, one going to each lung.
- The bronchi divide into smaller and smaller tubes.
- Finally they end in the air sacs called alveoli.

## Gaseous exchange

It is in the alveoli that **gas exchange** occurs. They are effective exchange surfaces. In the **alveolus**:

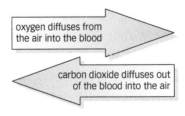

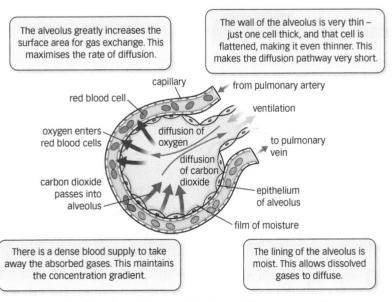

The alveolus greatly increases the surface area for gas exchange. This maximises the rate of diffusion.

The wall of the alveolus is very thin – just one cell thick, and that cell is flattened, making it even thinner. This makes the diffusion pathway very short.

There is a dense blood supply to take away the absorbed gases. This maintains the concentration gradient.

The lining of the alveolus is moist. This allows dissolved gases to diffuse.

▲ How the alveoli and capillaries in the lungs aid gaseous exchange.

# Ventilation

**Ventilation** is the movement of air into (inhaling) and out of (exhaling) the lungs.

## Breathing in – inhaling

1   The intercostal muscles between the ribs contract, lifting the rib cage up and out, expanding the thorax.
2   The diaphragm muscle contracts, flattening the diaphragm. This also expands the thorax.
3   The volume inside the lungs increases, and the pressure decreases.
4   Air rushes into the lungs due to the low pressure.

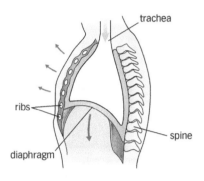

## Breathing out – exhaling

1   The intercostal muscles relax, and the ribs move down and in, reducing the volume of the thorax.
2   The diaphragm muscle relaxes, and arches up.
3   The volume inside the lungs decreases, which increases the pressure in the lungs.
4   The higher pressure forces air out of the lungs.

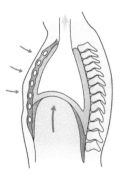

Some people experience difficulty in breathing. Asthmatics suffer from restriction of the bronchioles, which they treat with inhalers. These contain a drug to relax the bronchioles, which allows air to be inhaled and exhaled more freely. If ventilation stops, its action can be taken over by artificial ventilators to prevent damage to the organs of the body through lack of oxygen, and of course to preserve life.

## Exam tip

You should be able to recognise and label the major organs of the respiratory system on a diagram.

## Questions

1   What is the function of the diaphragm?

2   Which muscles move the ribs?

3   **H** What happens to the oxygen we breathe in?

## Plant exchange surfaces

Plants also need exchange surfaces. There are two major exchange surfaces in plants:

- leaves – where water vapour, carbon dioxide, and oxygen are exchanged with the air by diffusion
- roots – where water and minerals ions are absorbed.

## Gas exchange in the leaf

The leaf is efficiently designed as an exchange surface:

- It has a flattened shape, giving a large surface area for **gas exchange**.
- It has many internal air spaces, which again increase the surface area for exchange.
- The lower surface has a large number of **stomata**. These are small pores in the leaf, protected by a pair of guard cells, which open to allow molecules to move into and out of the leaf.

Three gases move into or out of the leaf when the stomata are open.

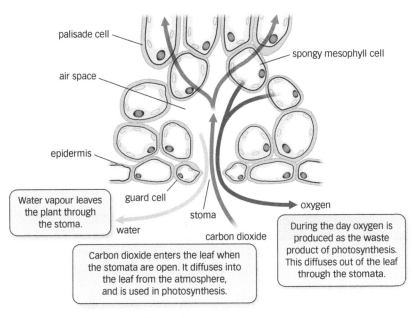

Water vapour leaves the plant through the stoma.

Carbon dioxide enters the leaf when the stomata are open. It diffuses into the leaf from the atmosphere, and is used in photosynthesis.

During the day oxygen is produced as the waste product of photosynthesis. This diffuses out of the leaf through the stomata.

▲ Gaseous exchange in the leaf.

## The exchange of water

Water and mineral ions move into plants through the roots. The roots are adapted for exchange because they have root hairs, which increase the surface area for absorption.

### Revision objectives

- ✓ know how plants exchange gases through their leaves, and the role of the stoma
- ✓ understand how water is absorbed and lost in transpiration
- ✓ explain the effects of environmental factors on the rate of transpiration

### Student book references

3.6   Exchange in plants

3.7   Rates of transpiration

### Specification key

✓ B3.1.3

water **evaporates** from the leaves – this process is called **transpiration**

water moves from the roots to the leaves – this is called the **transpiration stream**

water is absorbed by the root hairs

▲ The transpiration of water through a plant.

## Controlling water loss

As well as allowing gaseous exchange, the stomata on a leaf control water loss.

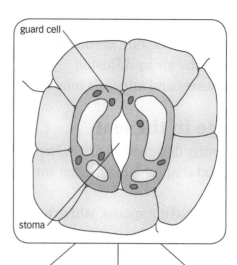

| Most stomata are on the lower surface where there is less heat from the Sun, to reduce water loss. | The guard cells can change shape to open and close the stoma. Stomata are usually open in the day to allow the gases in for photosynthesis, but closed at night to reduce water loss. | When the plants lose water faster than it is replaced from the roots, the stomata close to reduce water loss and to stop the plant wilting. |

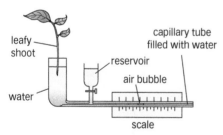

▲ The **potometer** is the piece of apparatus used to measure the rate of transpiration.

**Questions**

1   What is the role of the stomata?

2   List **three** factors that affect the rate of transpiration.

## Factors that affect the rate of transpiration

There are four main factors that affect the **rate of transpiration**.

| Factor | Effect on rate of transpiration |
|---|---|
| Increased light intensity | Increases – this is because stomata open in the light. As more stomata open, more water can evaporate, so transpiration increases. Eventually all the stomata open and the rate plateaus. |
| Increased temperature | Increases – the higher the temperature, the faster the water molecules in the air will move. This means they evaporate from the leaf quicker. |
| Increased air movement | Increases – when air moves over the leaf, it moves evaporated water away from it. The faster the air movement, the quicker the water is moved away, increasing the concentration gradient. |
| Increased humidity | Decreases – because if there are more water molecules in the air, the concentration gradient between the outside and inside of the leaf is reduced. It will take longer for water molecules to diffuse out of the leaf. |

### Working to Grade E

1 What happens to diffusion as the organism gets bigger?
2 Organisms have special exchange systems.
   a Why do organisms need such systems?
   b Give **three** examples of such systems, including at least **one** plant and **one** animal system.
3 Where are the lungs located?
4 The ribs surround the lungs. What **two** functions do the ribs have?
5 What **two** gases are exchanged across the lungs?
6 In which direction do the ribs move when we breathe out?
7 Where does gas exchange occur in the lungs?
8 Name the muscular sheet below the lungs, which separates them from the abdomen.
9 What is ventilation?
10 When we breathe in, the air passes through a sequence of structures. Put the following structures in the correct order.

| bronchus   alveolus   trachea   mouth |
| --- |

11 Look at the diagram of the respiratory system below. Label parts A–E.

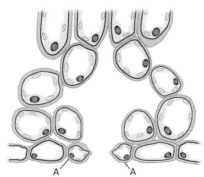

12 What are the **two** major exchange surfaces in the plant?
13 What is the transpiration stream?
14 Below is a drawing of a stoma shown in a section through the leaf.

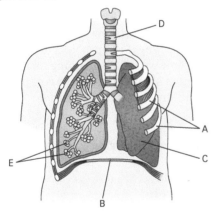

   a Label cells A on each side of the stoma.

   b Using arrows, show on the drawing the direction of movement through the stoma of:
      i oxygen
      ii carbon dioxide.
15 What piece of apparatus is used to measure the rate of transpiration in plants?

### Working to Grade C

16 All exchange systems are modified to perform their functions.
   a Complete the table below, which shows how such systems are adapted to their functions.

| Feature | How it improves diffusion |
| --- | --- |
| large surface area | |
| thin surface | |
| | to maintain a concentration gradient |

   b Look at the diagram of the human villus below. Identify **three** features of the villus that are adaptations for efficient exchange.

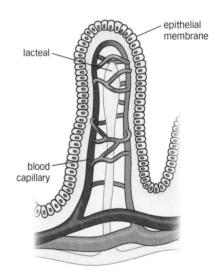

17 Explain why small organisms do not need an exchange system.
18 Give **three** adaptations of the lungs that make them an efficient exchange surface.
19 What causes the pressure to increase inside the lungs when we breathe out?

**20** Complete the following table to say whether a gas has increased, decreased or stayed the same, in the air we breathe out, compared to the air we breathe in.

| Gas | Change |
|---|---|
| oxygen | |
| carbon dioxide | |
| water vapour | |
| nitrogen | |

**21** List **three** factors that affect the rate of transpiration.

**22** Stomata are the pores through which gas exchange occurs in the leaf.
   **a** On which surface of the leaf are stomata mainly found?
   **b** Explain why this is an advantage.

**23** Look back at question 14.
   **a** Using an arrow, show on the drawing the direction of movement through the stoma of water.
   **b** When does the stoma mainly open?

**24** Explain why it is an advantage for most leaves to have a large surface area.

**25** Describe **one** way a root is adapted to be efficient at diffusion.

**26** What happens to a plant that is unable to replace water as quickly as it is lost?

## Working to Grade A*

**27** Think about the features of the villus you identified in question 16b. Explain how these features make diffusion more efficient.

**28** Below is a drawing of two cube-shaped cells of different sizes. The lengths of the sides of the cell are shown.

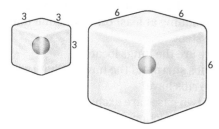

   **a** Complete the table below by calculating:
     **i** the surface area of the cells
     **ii** the volume of the cells
     **iii** the surface-area-to-volume ratio of the cells.

| Cell dimensions | Surface area | Volume | Surface-area-to-volume ratio |
|---|---|---|---|
| 3 | | | |
| 6 | | | |

   **b** What do you notice about the size of the ratio as the cell gets larger?
   **c** Explain why it is a problem to be big.

**29** Explain how we inhale. In doing this your explanation should include the role of the following structures in the process:
   • ribs
   • diaphragm.
You should also describe what happens inside the lung to:
   • the volume
   • the pressure.

**30** Most leaves have a large surface area.
Plants like Marram grass, which live on dry sand dunes, have reduced the surface area of their leaves. Explain why.

**31** Describe the relationship between the rate of transpiration and the following factors:
   **a** light intensity
   **b** air movements
   **c** humidity

# Examination questions
# Exchange surfaces in the lungs and in plants

**1** Two students decide to investigate breathing rates.

One student acts as the subject and the other records the number of breaths taken in 20 minutes by the other as 320.

**a** Calculate the breathing rate per minute. Show your working in the space below.

..........................................................................................

.................... breaths per minute.
*(2 marks)*

**b** Predict what might happen to the breathing rate of the student if they begin to exercise.

..........................................................................................
*(1 mark)*

**c** Gas exchange takes place in the alveolus.

**i** What gas does the student remove from the air?

..........................................................................................
*(1 mark)*

**ii** Describe how the alveolus is adapted to be an efficient exchange surface.

..........................................................................................

..........................................................................................

..........................................................................................

..........................................................................................

..........................................................................................
*(3 marks)*
***(Total marks: 7)***

**2** A plant biologist designed an experiment to investigate the effect of environmental conditions on the rate of water loss in plants.

The biologist recorded the rate of water uptake by the plants in different conditions of light and wind. The graph shows the results.

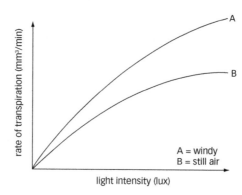

A = windy
B = still air

**a** Describe the relationship between the light intensity and the rate of water loss.

.................................................................................................................................................................

.................................................................................................................................................................

.................................................................................................................................................................
*(1 mark)*

**b** Water is lost through tiny pores called stomata on the leaf.

**i** What do you think happens to the stomata in the dark?

.................................................................................................................................................................

.................................................................................................................................................................
*(1 mark)*

**ii** What evidence allows you to think this?

.................................................................................................................................................................

.................................................................................................................................................................

.................................................................................................................................................................
*(1 mark)*

**c** Gardeners need to water their plants. Use the graph to explain in which weather conditions gardeners will need to water their plants most.

.................................................................................................................................................................

.................................................................................................................................................................

.................................................................................................................................................................

.................................................................................................................................................................
*(2 marks)*
***(Total marks: 5)***

## The need for a circulatory system

As animals get larger they need a circulatory system. This is because diffusion becomes too inefficient to move molecules like:

- oxygen – from the surface, deep into the cells of the body
- waste carbon dioxide – from the cells, to the outside of the body
- foods – from the small intestine, to the cells of the body.

Circulatory systems transport these substances around the body.

## Parts of a circulatory system

Human circulatory systems have three component parts:

- Blood – a fluid to carry the molecules.
- The **heart** – a pump to move the blood around the body.
- Vessels – tubes to contain the blood.

## The human circulatory system

Humans have a double circulatory system. This means that the blood passes through the heart twice as it makes its way around the body. The heart pumps deoxygenated blood to the lungs in the first circuit, and oxygenated blood to the body in a second circuit.

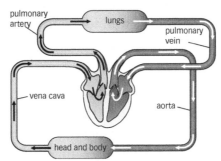

▲ The human double circulatory system.

In a complete **circulation** blood passes:

- from the heart to the lungs to remove carbon dioxide and collect oxygen
- back to the heart
- to the body organs and tissues
- back to the heart before going to the lungs again.

## The heart

The heart is an organ that pumps blood around the body. Typically the heart beats 60–80 times a minute. Much of the wall of the heart is made of muscle tissue. The heart is divided into four chambers (left and right atria and left and right ventricles). The atria have thin walls as they only pump blood to the ventricles. The ventricles have thick walls as they pump blood all around the body.

## Revision objectives

- ✔ know that substances are transported around the body in the circulatory system
- ✔ understand there are two separate circulations: one to the lungs, and one to the other organs of the body
- ✔ know the structure of the heart and the circulation of blood through the heart

## Student book references

**3.8** The circulatory system

**3.9** The heart

## Specification key

✔ B3.2.1 a – d

▼ Section through the human heart showing circulation of the blood.

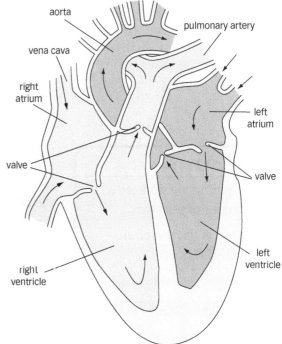

## Circulation through the heart

1   Deoxygenated blood arrives from the body through the **vena cava** to the right atrium.
2   The right atrium contracts and forces the blood into the right ventricle.
3   The right ventricle contracts and forces the blood up and out of the heart through the **pulmonary artery**.
4   There is a **valve** between the ventricles and atria, which is forced shut when the ventricles contract, preventing backflow of blood, so the blood flows in the right direction.
5   A second valve prevents blood from the artery draining back into the heart.
6   Blood goes to the lungs and picks up oxygen and loses carbon dioxide.
7   Oxygenated blood returns to the left atrium of the heart through the **pulmonary vein**.
8   The atrium contracts and forces blood into the left ventricle.
9   The left ventricle contracts and forces the blood out of the heart through the **aorta**, to the body. The left ventricle has a thicker wall to pump the blood all around the body.
10  The two valves again prevent the backflow of blood in the heart.

These two processes happen at the same time: the atria contract together, then the ventricles contract together, so the process of blood moving round your body and through your lungs is a continuous flow.

## Artificial hearts and heart valves

Common circulatory conditions include:

- Heart failure – where the heart muscle is damaged, and struggles to contract.
- Valve damage – where backflow of blood is not prevented.

Developments in biomedical and technological research enable us to help treat these conditions.

Artificial hearts have been developed. These can be used as a short-term solution to heart failure whilst waiting for a transplant. These keep the patient alive and are not rejected by the body. However, they often have wires that protrude through the skin. It is hoped that artificial hearts will improve.

Artificial valves are plastic and metal valves that can replace damaged valves. These are long lasting and highly successful.

## Exam tip

Make sure you can identify and label the four chambers of the heart and the names of the blood vessels associated with them. Learn the circulation of blood through the heart as a sequence. You could learn them as series of numbers, or colour each statement with a sequence of colours.

## Questions

1   Explain why we need a circulatory system.

2   What is the use of an artificial heart?

3   **H** How does the heart pump blood around the body?

## The blood vessels

The blood vessels are the tubes through which the blood flows. There are three types of blood vessel.

**Arteries** take blood away from the heart, **capillaries** take blood through the organs, and **veins** return blood to the heart. Their structure is related to their functions.

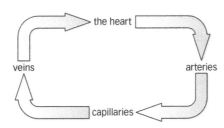

### Arteries

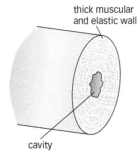

thick muscular and elastic wall

cavity

- walls are thick because the blood is under high pressure
- large amounts of **muscle** allow the wall to withstand and maintain the pressure
- large amounts of **elastic fibres** allow the artery to stretch and recoil as blood surges through
- narrow cavity or lumen

### Capillaries

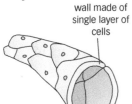

wall made of single layer of cells

- walls are very thin, only one cell thick, so diffusion is quick
- large number of capillaries gives a large surface area for diffusion
- molecules needed by the cells (such as oxygen and glucose) pass out of the blood
- molecules produced by the cells (carbon dioxide and wastes) pass into the blood
- blood pressure has been lost, and the blood flows slowly by the time the blood reaches the capillaries
- very narrow, just wide enough to allow one red blood cell through

### Veins

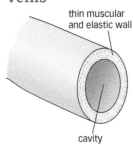

thin muscular and elastic wall

cavity

- thinner walls than arteries, because the blood pressure is lower
- little muscle or elastic fibres as there is no high pressure to withstand
- valves to prevent backflow of blood
- large lumen

### Revision objectives

- ✔ be able to describe the structure and function of arteries, veins, and capillaries
- ✔ understand the use of stents
- ✔ understand the composition of the blood and the function of each part

### Student book references

3.8   The circulatory system

3.10 Arteries and veins

3.11 Capillaries

3.12 Blood

3.13 Blood cells

### Specification key

- ✔ B3.2.1 e – g
- ✔ B3.2.2

## Treating narrowed arteries

A common circulatory problem is narrowing of the arteries. This is usually due to the build-up of fat in the wall of the artery. If this occurs in the arteries supplying blood to the heart muscle (the coronary arteries), it could result in a heart attack. To treat this condition a wire mesh is inserted into the narrow region of the artery and expanded. This **stent** opens the artery.

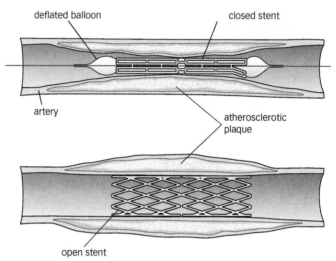

▲ A coronary stent.

# Blood

Blood is a tissue, because it is made of similar cells working together. It is a fluid that flows through the blood vessels, pumped by the heart. It has three main functions:

- Transport – carries substances and cells around the body.
- Protection – from infection and blood loss.
- Regulation – helps to maintain the body temperature and pH.

## Composition of the blood

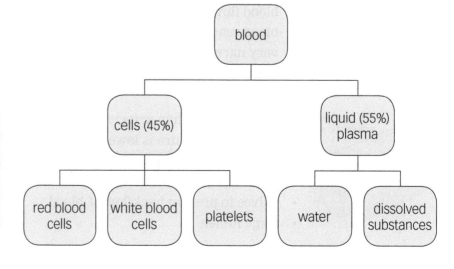

## A closer look at the blood

| Component | Function |
|---|---|
| **Plasma** | Transports dissolved substances such as:<br>• carbon dioxide from the cells to the lungs<br>• soluble products of digested foods from the small intestine to the rest of the body<br>• urea from the liver to the kidneys. |
| Red blood cell | • Contains the red pigment called **haemoglobin**. This combines with oxygen in the lungs to form oxyhaemoglobin.<br>• Red blood cells transport the oxygen around the body; oxyhaemoglobin then breaks up to release the oxygen in other organs.<br>• There is no nucleus, which provides more room for haemoglobin.<br>• Made in the bone marrow, and destroyed in the liver. |
| White blood cell<br>nucleus | • There are several types. They all contain a nucleus. They all form part of the immune system, working to fight infection.<br>• Some engulf and digest microorganisms, others make antibodies to destroy microorganisms. |
| Platelets | • Small fragments of cells with no nucleus. They help form blood clots at the site of wounds, to prevent blood loss and infection. |

## Key words

artery, capillary, vein, muscle, elastic fibre, stent, plasma, haemoglobin

## Artificial blood products

It is important to maintain our blood volume. When people have serious injuries, such as during war or major trauma, it can result in major blood loss. It is important to replace lost volume. When real blood is not available, artificial blood can be used. This will not be rejected by the body and will maintain blood pressure. Some types also contain chemicals to help transport oxygen.

## Questions

1  Does an artery take blood to or away from the heart?

2  Why does an artery have a thicker wall than a vein?

3  **H** Explain how the blood helps to protect us against disease.

## Revision objectives

- ✓ know the role of xylem in the transport of water and ions in plants
- ✓ know the role of phloem in the transport of sugars in plants

## Student book references

3.14 Transport in plants

## Specification key

- ✓ B3.2.3

## Key words

xylem, phloem, transpiration stream, translocation

## Exam tip

**AQA**

This topic is about two tissue types with two separate functions: learn the table of comparisons above. Look back at transpiration in plants, and link the ideas to this spread.

## Questions

1 How does the plant transport water?

2 How does the plant transport sugars?

3 **H** Where is sugar transported from and to?

# Plant transport systems

Plants also have transport systems. The separate systems involve two types of tissue called:

- xylem
- phloem.

They are located in a vascular bundle.

## Summary of transport in plants

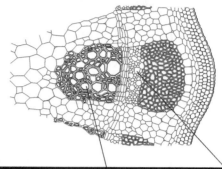

| Tissue name | Xylem | Phloem |
|---|---|---|
| What does the system transport? | water and mineral ions | dissolved sugars |
| From where? | roots | leaves (source) |
| To where? | leaves | rest of plant (sink), including growing regions and storage organs |
| Cell structure | dead cells stacked on top of each other to form tubes | living cells stacked on top of each other to form tubes |
| Name of process | transpiration stream | translocation |

## A closer look at the transpiration stream

The diagram shows the detail of how water and minerals are transported from the root to the leaves.

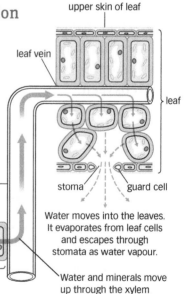

upper skin of leaf

leaf vein

leaf

stoma / guard cell

Water moves into the leaves. It evaporates from leaf cells and escapes through stomata as water vapour.

Water and minerals move up through the xylem vessels to the stem and the leaves

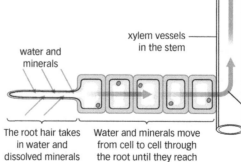

water and minerals

xylem vessels in the stem

The root hair takes in water and dissolved minerals from the soil

Water and minerals move from cell to cell through the root until they reach xylem vessels

▲ The process of transpiration.

# Questions
## Transport in animals and plants

### Working to Grade E

1 What are the **three** major parts of the human circulatory system?
2 What gas does the blood pick up at the lungs?
3 Below is a diagram of the human heart.

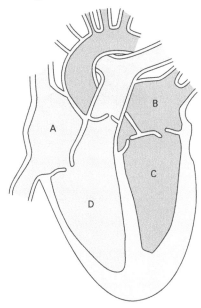

   a Name the four chambers labelled A–D.
   b Label the aorta.
   c Label a valve.
4 Which type of blood vessel takes blood away from the heart?
5 Which type of blood vessel takes blood to the heart?
6 Name **two** substances transported in the plasma.
7 The blood contains a number of cells.
   a What is the function of the red blood cell?
   b What is the name of the pigment that helps the red blood cell carry out its function?
8 What is transported in the xylem?
9 What is transported in the phloem?

### Working to Grade C

10 Why is the human circulatory system described as a double circulation?
11 What is the name of the blood vessels which take blood:
   a to the lungs?
   b to the heart from the body?
12 Look at the diagram in question 3. Use arrows to show the circulation of deoxygenated blood through the heart.
13 What is the function of the valves in the heart?
14 The heart chambers are surrounded by walls of muscle.
   a Why are the walls of the ventricles thicker than the walls of the atria?
   b Why is the wall of the left ventricle thicker than the right ventricle?

15 How would a surgeon treat a heart with a damaged valve?
16 Artificial hearts have been developed to treat heart failure.
   a Give an advantage of using an artificial heart.
   b Give a disadvantage of using an artificial heart.
17 The blood contains a number of cells.
   a What type of cell produces antibodies?
   b What is the function of a platelet?
   c Why do red cells have no nucleus?
18 What are the **three** functions of the blood?
19 Artificial blood has now been developed. Give **one** situation where artificial blood might be used.
20 Stents are used to treat some circulatory problems. What is a stent?
21 Name a difference between xylem and phloem tissues.
22 What is translocation?
23 In translocation biologists often talk about substances moving from a source to a sink.
   a What do we mean by the term 'source'?
   b Give an example of a source.
   c Give an example of a sink.

### Working to Grade A*

24 Arteries and veins are two types of blood vessel.
   a List **three** differences between arteries and veins.
   b Explain the reason for the differences.
25 Explain why the walls of capillaries are so thin.
26 Give an advantage of using artificial blood.
27 Stents are used to treat some circulatory problems. Describe how they work.
28 Describe the path taken by water molecules as they move through the plant in the transpiration stream.

**1** Below is a drawing of human blood cells.

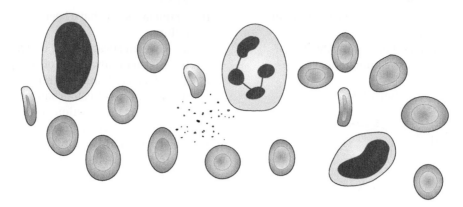

**a** What is the function of the red blood cell?

.................................................................................................................................................................

.................................................................................................................................................................

*(1 mark)*

**b** How do the platelets help prevent infection?

.................................................................................................................................................................

.................................................................................................................................................................

*(1 mark)*

**c** How do the white blood cells help deal with infection?

.................................................................................................................................................................

.................................................................................................................................................................

.................................................................................................................................................................

*(2 marks)*
**(Total marks: 4)**

**2** The heart pumps blood around the body. The heart contains valves. Sometimes the valves are damaged by disease, and fail to work. This may lead to heart failure or heart attacks.

**a** What is the function of the valves in the heart?

.................................................................................................................................................................

.................................................................................................................................................................

.................................................................................................................................................................

*(1 mark)*

**b** One treatment for damaged valves, which has been developed since the 1950s, is to replace them with an artificial valve.

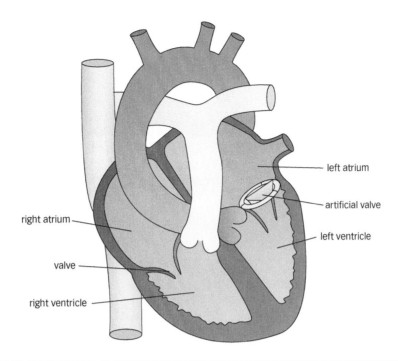

- • Damaged heart valves can be replaced by open-heart surgery.
- • Once fitted the valves last up to 30 years.
- • Valves can cause damage to blood cells, which results in blood clotting, so patients must take anticlotting drugs for the rest of their lives.

Suggest advantages and disadvantages of treating patients with artificial valves.

..................................................................................................................................................................

..................................................................................................................................................................

..................................................................................................................................................................

..................................................................................................................................................................

..................................................................................................................................................................

..................................................................................................................................................................

..................................................................................................................................................................

..................................................................................................................................................................

*(4 marks)*
***(Total marks: 5)***

**3** Look at the drawing of the human heart.

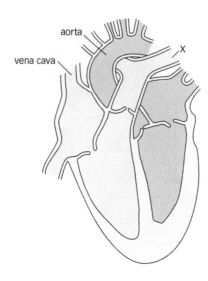

aorta

vena cava

X

**a** What is the name of the blood vessel labelled X?

.......................................................................................................................................................................

*(1 mark)*

**b** To which organ does blood vessel X take the blood?

.......................................................................................................................................................................

*(1 mark)*

**c** Describe the route taken by the blood as it moves through the heart from the vena cava to the aorta.

.......................................................................................................................................................................

.......................................................................................................................................................................

.......................................................................................................................................................................

.......................................................................................................................................................................

.......................................................................................................................................................................

.......................................................................................................................................................................

.......................................................................................................................................................................

.......................................................................................................................................................................

.......................................................................................................................................................................

.......................................................................................................................................................................

.......................................................................................................................................................................

.......................................................................................................................................................................

.......................................................................................................................................................................

*(8 marks)*

***(Total marks: 10)***

## Research and development – ongoing review of ideas

In the exam you may be given information to read and then be asked to answer questions using that information and your own knowledge. The topics can vary but new technologies that can have a direct impact on our way of life are often chosen.

One such topic is the development of artificial aids to breathing.

## Invention – the iron lungs of the 1920s

1   Read the information below. Using this and your own knowledge, answer the questions that follow.

We have learnt about the importance of ventilation to maintaining life. Before the 1920s any condition that paralysed the diaphragm and intercostal muscles would stop natural ventilation and the patient would die. This could be caused by exposure to some gases or the disease polio. In 1929 Dr Robert Henderson invented the iron lung to treat these patients.

Although the iron lung looks primitive, at the time of its invention it must have seemed like a miracle. Its first use was on an unconscious child, who responded rapidly to the treatment and woke up. By the 1950s when there was a polio epidemic, hospital wards filled with iron lungs were not uncommon, despite their high cost.

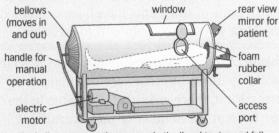

The bellows causes air pressure in the 'lung' to rise and fall. The patient's lungs are aided in drawing in air.

For the duration of their illness, the person is placed in a sealed chamber, with only their head exposed. Air is drawn out of the chamber, causing the pressure to drop in the chamber, and the chest to expand, drawing air into the lungs.

### Skill – applying knowledge to an innovative solution

a   The iron lung replaces the actions of the body.
  i   Which parts of the body does it replicate?
  ii  Give one way they are similar and one way they are different.

Here you need to connect your existing knowledge to an unfamiliar situation. The connection is to explain how the body draws air into the lung. Then you explain that in the body the ribs and diaphragm carry out the function of the pressure change in the chamber. The similarities are often that the same outcome is achieved, but by a different mechanism.

### Skill – evaluating the developments in science and technology

b   Suggest the advantages and disadvantages of treating patients with the iron lung.

When given a question that asks you to evaluate the advantages of a medical development, the advantage should often relate to prolonging life.

In contrast, when trying to list the disadvantages they often fall into one of two categories:

**Social or financial**
Here the text refers to the cost, and also that the body is enclosed in the machine, making it difficult for a doctor to treat the patient.

**Ethical**
This often means the quality of life; here it would be limited as patients would be in the machine for a long time.

## Progression – modern ventilators

2   Now read the information below and answer the question that follows.

As iron lungs became more commonly used, doctors became frustrated by their limitations. By the 1950s they had developed a 'positive pressure ventilator'. This device did not place the patient in a large sealed chamber. Instead a pipe can be placed down into the lungs. Air is then gently pumped into the lungs. The pump uses positive pressure to expand the lungs. In long-term cases the pipe can be inserted into the windpipe, or trachea, at the neck.

### Skill – Explaining technological developments

a   Explain how the positive pressure ventilators are a medical advancement.

Here you will be expected to apply your knowledge to this situation. You start from the problems identified for the iron lung, and look at the advancements implied in the text about the positive pressure ventilator. They include:
- allowing patients more freedom of movement
- allowing the doctor access to the patient's body for examination and surgery
- that they could be used to ventilate any patient during open-chest surgery, when the ribs don't work.

### Moving forward – completing the story

Doctors are still working to improve the design of ventilators. Modern ventilators can detect whether the patient is making any attempt to breathe naturally. The device detects changes in pressure inside the lung caused by the patient, and responds by gradually phasing down its action, as the patient recovers. This would have been unthinkable in the iron lung of the last century.

In recent years scientists have developed artificial blood. This can be used in medical treatments requiring blood transfusions, but is particularly useful on battlefields. Below is a table of information comparing artificial blood to human blood.

| Characteristic | Artificial blood | Human blood |
|---|---|---|
| **Storage** | room temperature | refrigerated |
| **Shelf life** | up to 36 months | up to 42 days |
| **Active life** | 1–2 days | several weeks |
| **Compatability** | no antigens, so can be used in all patients | antigens present, needs tissue matching |
| **Oxygen carriage** | effective | decreases with storage time |
| **Purity** | pathogen free | needs to be screened for pathogens |
| **Complications** | increased risk of heart disease and heart attack | none |

Suggest advantages and disadvantages of using artificial blood. *(5 marks)*

Artificial blood is easy to store at ...

**Examiner:** Here the student has made several errors. The ...

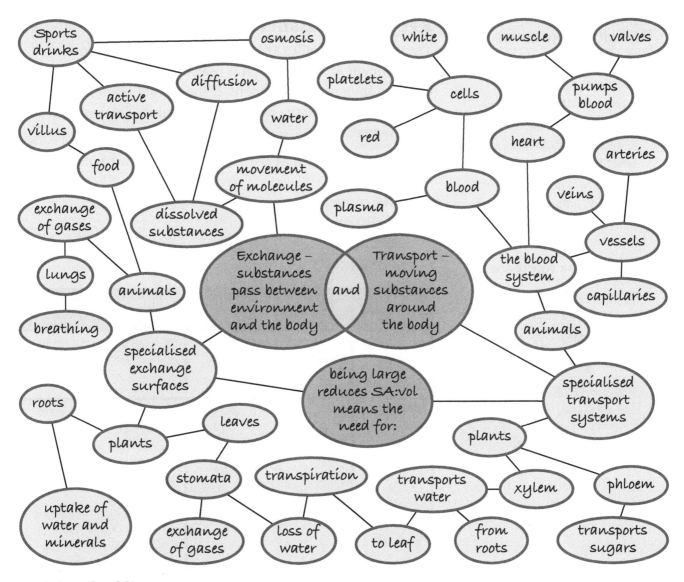

## Revision checklist

- Dissolved substances can move into or out of cells by diffusion or active transport. Water moves into and out of cells by a special type of diffusion called osmosis.
- Keeping the body hydrated is important for human health. Water lost in sweat during exercise must be replaced.
- Sports drinks contain water for hydration, but also glucose for energy and ions to keep the body healthy.
- Active transport moves substances across cell membranes against a concentration gradient. This requires energy and a protein carrier.
- As organisms get bigger their surface-area-to-volume ratio gets smaller. This makes diffusion inefficient. To solve the problem they develop special exchange surfaces.
- All exchange surfaces are thin, have a large surface area, and have systems to maintain a concentration gradient.
- The villus is the site of absorption in the small intestine, and is efficiently designed.
- The lungs are the organs of gas exchange in humans. Breathing is the process which brings air in and out of the lungs.
- Transpiration is the loss of water from the leaves of the plant. Water leaves the plant through tiny pores called stomata on the under surface of the leaf.

- The rate of transpiration can be measured by a potometer, and can be affected by a number of environmental factors.
- A second problem for larger organisms is that they require a transport system to move substances around their body.
- Humans have a circulatory system that involves the heart, blood, and blood vessels.
- The heart is a muscular pump that pushes the blood around the body. Blood passes through the heart twice in one cycle of the body; it is a double circulatory system.
- The blood is the fluid that transports substances. It is composed of a liquid called plasma, and cells – red cells, white cells, and platelets – each with their own function.
- The blood is circulated in vessels: arteries take blood away from the heart, capillaries take blood through the tissues, where exchange occurs, and veins take blood back to the heart.
- Technology has developed artificial blood for use in transfusions, and stents to open blocked vessels. Artificial hearts or valves can be used to replace damaged ones.
- Plants also transport substances. The xylem transports water from roots to the leaves, whilst the phloem transports sugars from the leaf to other parts of the plant.

## Revision objectives

- ✓ understand that homeostasis is maintaining a constant internal environment
- ✓ understand that excretion is the removal of toxic substances from the body
- ✓ know how a healthy kidney produces urine

## Student book references

**3.15** Keeping internal conditions constant

**3.16** The kidney

## Specification key

✓ B3.3.1 a – c

# Keeping it balanced

The cells of our bodies are sensitive to any changes in the environment around them. It is essential to maintain a constant internal environment. If these conditions vary they are brought back to the body's normal value. This is called **homeostasis**.

Internal conditions that are kept within narrow limits are:

- pH
- **water** content
- ion (salt) content
- temperature
- blood **sugar** levels.

| Internal condition | Effect on the body | Homeostatic function |
|---|---|---|
| pH | Chemicals like **carbon dioxide** or lactic acid lower the blood pH, damaging proteins. | The lungs control the levels by altering the breathing rate. |
| Water and ions | Water and ions are taken into the body in food and drink. If the content of these in the body is wrong, it affects osmosis. Water may move into or out of our cells, both resulting in damage. | The **kidneys** vary the levels in the urine.<br><br>When we have drunk a lot of water, the kidneys produce lots of dilute urine, removing the water.<br><br>When we are dehydrated the kidneys produce a little concentrated urine, thus saving water. |
| Temperature | Human body temperature is kept at 37 °C. It will rise and fall, which will affect cell function. | In the skin heat is lost or conserved. |
| Blood sugar levels | Sugar is taken in through our diet and used in respiration. Excess is stored. If the levels are too high or low then the body becomes ill. | The pancreas produces hormones that cause the sugar to be stored or released. |

Many of the reactions in the cells of our body also produce toxic wastes, which must be removed. This process is called **excretion**.

Wastes that must be removed include:

- carbon dioxide
- **urea**.

| Waste | Production | Excretion function |
|---|---|---|
| Carbon dioxide | Carbon dioxide is a waste product of respiration. A build-up will lower pH, and damage proteins in the cell. | Breathed out through the lungs. |
| Urea | Produced in the liver by the breakdown of excess amino acids from proteins. This is toxic to cells because it is alkaline. | Removed by the kidneys, and stored in the bladder until urination. |

## The kidney

The kidneys are the organs that remove wastes like urea, excess water, and ions. This is done by producing urine in the kidney. This trickles down thin tubes and is stored in an organ called the bladder. When the bladder is full we urinate and release the urine to the outside of our body.

### How does a kidney work?

Zone of filtration – here blood is **filtered** and wastes like urea are removed from the blood. Unfortunately small useful substances are also removed. Large proteins and blood cells are too large to be filtered.

Zone of **re-absorption** – here the small useful molecules are taken back into the blood by active transport. The molecules include sugar, dissolved ions, and as much water as the body needs to maintain the correct level of hydration.

Blood arrives at the kidney containing wastes

Filtered blood leaves the kidney

The final product is called urine. This trickles down this tube to be stored in the bladder.

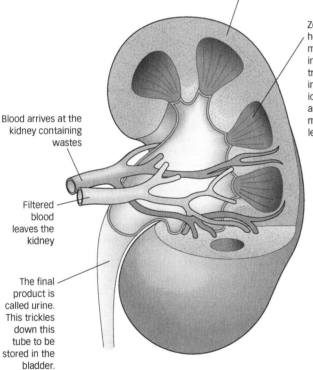

▲ The structure of the kidney.

## Questions

1  Why do we need to keep internal conditions constant?

2  What conditions does the body control?

3  **H** Explain how the kidney gets rid of waste.

## Revision objectives

- understand how kidney dialysis works
- understand the process of kidney transplant and tissue typing
- be able to evaluate the two treatments

## Student book references

**3.17** Renal dialysis

**3.18** Kidney transplants

## Specification key

✓ B3.3.1 d – i

# Kidney failure

Kidneys are important organs because they remove toxic wastes from our bodies such as urea. If the kidneys stop filtering out these toxins, it can make us very ill. This is called kidney failure. This can happen in a number of ways.

*Acute kidney failure* – here the kidneys stop working suddenly as a result of disease or drugs, but will recover. Treatment is by **dialysis**.

*Chronic kidney failure* – here the kidneys gradually fail to work, and do not recover. This could result from overuse of a drug, diabetes, or genetic causes. Treatment is either by dialysis or transplant.

# Kidney dialysis

The aim of dialysis is to remove the waste products from the blood, and restore the concentrations of all dissolved substances in the blood, like salts, to normal.

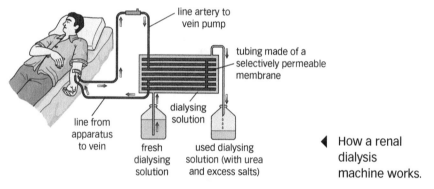

◀ How a renal dialysis machine works.

To achieve this, dialysis is usually carried out for five to six hours, three or four times a week.

The patient is attached to a dialysis machine, and their blood is taken from a vein and flows through the machine to be filtered.

In the dialysis machine the waste products are filtered out by diffusion.

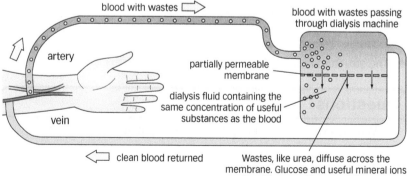

▲ How dialysis works.

# Kidney transplants

For patients with long-term kidney failure due to diseased kidneys, dialysis limits their quality of life. A transplant may be a better option. Here the diseased kidney is removed, and replaced with a healthy kidney. The healthy kidney is taken from a **donor** who might be a close relative or someone who has recently died. Care is taken to prevent **rejection** of the kidney by the **recipient's** immune system.

## Organ rejection

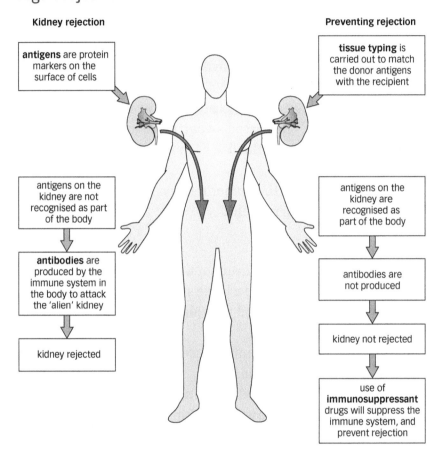

**Kidney rejection**

antigens are protein markers on the surface of cells

antigens on the kidney are not recognised as part of the body

**antibodies** are produced by the immune system in the body to attack the 'alien' kidney

kidney rejected

**Preventing rejection**

**tissue typing** is carried out to match the donor antigens with the recipient

antigens on the kidney are recognised as part of the body

antibodies are not produced

kidney not rejected

use of **immunosuppressant** drugs will suppress the immune system, and prevent rejection

## Evaluating treatments

| Treatment | Advantages | Disadvantages |
|---|---|---|
| dialysis | • effective waste removal<br>• allows time for the kidney to recover | • treatment time reduces quality of life<br>• expensive |
| transplant | • long-term solution<br>• better quality of life<br>• cheaper in the long term | • tissue matching<br>• lack of donors<br>• rejection |

## Exam tip

You need to be able to *evaluate* the types of treatment. Remember this means discussing both the advantages and disadvantages of the treatments.

## Questions

1 What is a kidney transplant?

2 How does kidney dialysis work?

3 **H** What causes organ rejection?

## Revision objectives

- ✔ be able to describe how thermoregulation is controlled
- ✔ explain the methods used in the body to bring the body temperature back to normal when it overheats
- ✔ explain the methods used in the body to bring the body temperature back to normal when it overcools

## Student book references

**3.19** Regulating body temperature – overheating

**3.20** Regulating body temperature – overcooling

## Specification key

✔ B3.3.2

# Body temperature

Our normal core body temperature is 37 °C. Our body tries to keep its temperature at this level, or very close to it, all the time. This is called **thermoregulation**. 37 °C is the temperature at which the body's organs work best. Sometimes our body will overheat or overcool depending on the environment or conditions in our body. Both situations are dangerous.

## Overheating

- Causes – being in an environment with high external temperature, exercise, or dehydration, which prevents sweating.
- Dangerous level – any temperature above 40 °C is dangerous.
- Body's response – increased **sweating** so more water needs to be taken in, looking flushed as more blood flows to the skin.
- Effect on the body – the higher temperature denatures enzymes, which harms cells.

## Overcooling

- Causes – being in an environment with a low external temperature; babies have a large surface-area-to-volume ratio, and so lose heat, and the elderly can find it difficult to maintain heat.
- Dangerous level – any temperature below 35 °C is dangerous, and is called hypothermia.
- Body's response – reduced sweating, **shivering** to generate heat, reduced **blood flow** to the skin, causing you to look paler.
- Effects on the body – the lower temperature slows enzyme reactions, making us unwell.

# The thermoregulatory centre

The changes in the body temperature are detected in two ways:
1  An area in the brain called the **thermoregulatory centre** monitors the temperature of the blood flowing through the brain.
2  Nerves in the skin detect skin temperature and send messages to the thermoregulatory centre.

If the temperature is too high or too low, the thermoregulatory centre will then trigger the body's response processes.

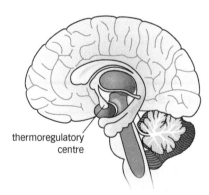

thermoregulatory centre

▲ The thermoregulatory centre (hypothalamus) in the brain.

# Mechanisms for thermoregulation

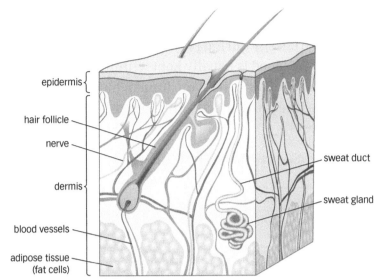

▲ Section through human skin showing sweat glands and sweat ducts.

Most of the mechanisms to help us control our body temperature involve the skin. It is important to be able to identify the major structures in the human skin.

| When you overheat | When you overcool |
|---|---|
| **AIM:** to cool the body down | **AIM:** to keep heat in the body |
| radiated heat from blood | |
| sweat gland    skin capillaries | sweat gland    skin capillaries |
| • Hair lie flat on skin; this reduces the layer of air, reducing insulation. | • Hairs stand up, trapping more insulating air. |
| **H** | |
| • Sweat glands release more sweat. This covers the skin and uses heat from the body to **evaporate** the water.<br>• Muscles do not cause shivering. | • Sweat glands release less sweat, so less heat is lost by evaporation.<br>• Muscles contract causing shivering. The contraction releases heat energy in respiration. |
| **Vasodilation**<br>• Blood vessels in the surface of the skin expand or dilate.<br>• This causes more blood to flow near the surface of the skin.<br>• More heat is lost from the skin by radiation. | **Vasoconstriction**<br>• Blood vessels in the surface of the skin get narrower, or constrict.<br>• This causes less blood to flow near the surface of the skin.<br>• Less heat is lost by radiation. |

## Questions

1 Name **one** thing that happens in the skin when we get too hot.

2 What is the thermoregulatory centre, and what does it do?

3 **H** What is vasodilation?

## Revision objectives

- ✓ understand the roles of the pancreas, insulin, and glucagon in the control of blood glucose levels
- ✓ know the causes and treatments of type 1 diabetes

## Student book references

**3.21** Regulating blood glucose levels

**3.22** Type 1 diabetes

## Specification key

✓ B3.3.3

# Regulating glucose in the blood

Glucose is an important molecule in our body. It is needed in cells to release energy during respiration. However, the level of glucose in the blood is usually kept within narrow limits. The normal level of glucose in the human blood is 72 mg per 100 cm$^3$. If the level falls or rises, the body takes action to regulate the level.

## How the body controls blood glucose

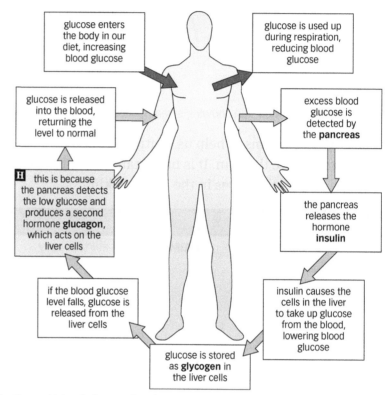

▲ Normal blood glucose level.

# Type 1 diabetes

Type 1 **diabetes** is a condition where the pancreas does not produce enough insulin to allow us to control (by lowering) our blood glucose.

## Key facts about type 1 diabetes

| | |
|---|---|
| **Frequency** | approximately 1 in 800 people |
| **Possible causes** | genetic; viral; some drugs; trauma |
| **Onset** | usually between 10 and 18 years old |
| **Symptoms** | thirst, frequent urinating |
| **How is it treated?** | • careful control of diet, reducing intake of sugary foods<br>• taking exercise<br>• monitoring blood glucose levels<br>• injecting insulin before a meal<br>• regular health checks on circulation and eyesight |

## Monitoring blood glucose levels

Blood glucose levels can be monitored by a simple digital recorder on a small sample of blood. This should be carried out at regular intervals. The graph on the right shows how the glucose levels would change after a meal, in both a normal and a diabetic patient.

## Monitoring blood insulin levels

Insulin levels can also be monitored, but this is carried out in hospitals. Below is a graph that shows the insulin levels after a meal, in both a normal and a diabetic patient.

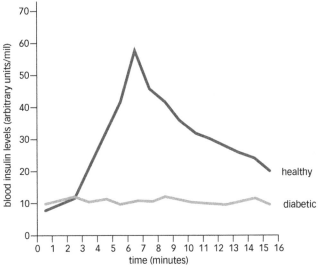

▲ Blood insulin concentrations in a normal and diabetic person immediately after a meal.

## Improvements in modern treatments

- Modern sensors to monitor blood glucose are simple and more effective.
- Human insulin produced by genetic engineering is now used to inject, so there is less risk of allergies.
- Circulation is checked more thoroughly, which reduces complications due to diabetes, such as gangrene.
- Automated insulin pumps that release insulin into the blood via a catheter have been developed. This means no injections are needed, and the patient can live a more normal life.
- Research is being carried out to develop stem cells that might replace the damaged pancreatic cells.
- Gene therapy research aimed at stopping the damage being caused to the insulin-producing cells is underway.

**Key words**

glucose, pancreas, insulin, glycogen, glucagon, diabetes

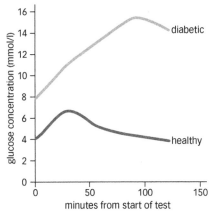

▲ Blood glucose concentrations in a normal and diabetic person immediately after a meal.

**Exam tip**

Don't muddle all the words here starting with the letter 'G'. Examiners call these the 'terrible G' words. Remember, *glucose* is the sugar, *glycogen* is the storage product, and *glucagon* is the hormone that breaks glycogen into glucose. Try to learn this, and their spellings.

**Questions**

1   Name **two** hormones involved in the control of blood glucose.

2   Which hormone is not produced by a type 1 diabetic?

3   **H** Explain how a low blood glucose level is restored to normal.

## Working to Grade E

1 Define homeostasis.
2 State one homeostatic function carried out by the following organs:
   a skin
   b kidney
   c pancreas
3 Define excretion.
4 What are the **two** major wastes produced in the body?
5 What is normal body temperature?
6 What is the function of the kidney?
7 Where are the kidneys located?
   • head
   • thorax
   • abdomen
8 What **two** types of treatment are used for kidney failure?
9 Name **two** causes of kidney failure.
10 What is the name given to the process of regulating body temperature?
11 Give **three** causes of the body overheating.
12 What is hypothermia?
13 Where is the thermoregulatory centre?
14 How does the body obtain glucose?
15 State **one** cause of type 1 diabetes.
16 What is glucose used for in the body?
17 Where is glucose stored in the body?
18 What is the effect of insulin on blood glucose levels?

## Working to Grade C

19 Below is a diagram of the human kidney.

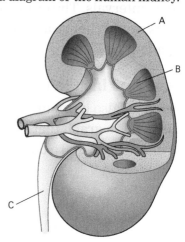

   a What happens in **A**?
   b Where does the waste product in **C** travel to?
20 Explain what causes the blood pH to fall.
21 Explain how urea is produced.
22 During dialysis describe what happens to the concentrations of urea in the blood.
23 What is an antigen?
24 During kidney transplant, who is the donor and who is the recipient?
25 The diagram shows the flow of blood through a dialysis machine.

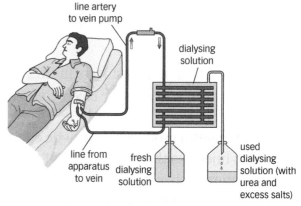

   a Draw an arrow labelled A to show the movement of urea during dialysis.
   b Label the partially permeable membrane.
   c Dialysis occurs frequently.
      i How long will a typical dialysis session last?
      ii Why does dialysis have to be repeated several times a week?
26 People often become flushed or red after doing exercise.
   a What causes this to happen?
   b What is the effect on the body?

**27** Below is a drawing of the human skin.

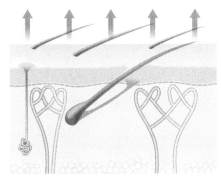

**a** State whether the diagram shows the skin on a hot day or a cold day.

**b** Give **three** reasons for your answer.

**28** How does the body monitor the external temperature?

**29** Why are body temperatures above 40°C dangerous?

**30** What regular healthcare checks are carried out on diabetics?

**31** What problems occur if glucose levels become too high in the body?

**32** Where is insulin made?

**33** Does type 1 diabetes usually develop in young people or old people?

**34** How has the development of an automated insulin pump improved quality of life for a diabetic?

**35** What are the symptoms of type 1 diabetes?

**36** Below are the results of regular blood glucose levels taken from two students, John and Peter.

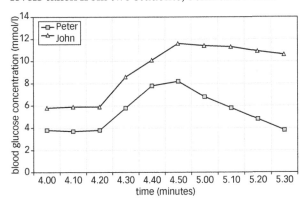

**a** The two boys had a sugary meal.
  **i** At what time do you think they had the meal?
  **ii** Give a reason for your answer.
**b** One of the boys is a diabetic.
  **i** Which boy is diabetic?
  **ii** Give a reason for your answer.
**c** Describe what will happen to the insulin levels during this time, in the blood of:
  **i** John
  **ii** Peter.
**d** What should the diabetic student do prior to a sugary meal?

**37** Look back at the diagram in question 20. Identify **B** and explain how it ensures that useful molecules are not lost from the body.

**38** Explain the effect of changes in the concentration of water or ions on the body.

**39** During exercise the body produces more carbon dioxide. Explain the effect of this on the breathing rate.

**40** Look back at the diagram in question 26.
  **a** Explain why the dialysis fluid is constantly being changed.
  **b** Explain how dialysis maintains and regulates the correct level of salts in a patient.

**41** Explain the difference between acute and chronic kidney failure.

**42** During kidney transplants organ rejection is a significant problem.
  **a** What is organ rejection?
  **b** How can we reduce the risk of organ rejection during transplants?

**43** Patients will often prefer a kidney transplant over dialysis. What scientific argument could you use to back up this point of view?

**44** Explain how sweating helps to regulate the body temperature.

**45** Explain why we need to drink more fluid on a hot day.

**46** Explain how hair could be used to increase **and** decrease the amount of insulation in an animal's body.

**47** Explain what vasoconstriction is, and its effects on temperature regulation.

**48** Explain the role of glucagon in regulating blood glucose levels.

**49** Evaluate how modern developments have improved the treatment of diabetes.

# Homeostasis

**1** It is important for humans to maintain their water balance. Look at the diagram, which shows the input and output of water for a human in a typical day.

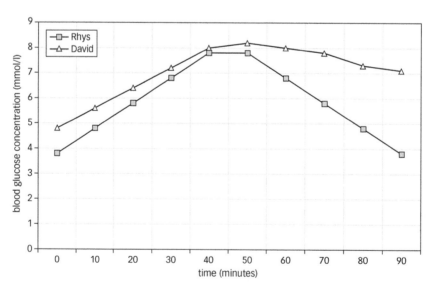

**a** To keep the water levels in the human in balance, what is the minimum amount of water this human would need to drink in a day?

.................................................................................................

.................................................................................................

........................ cm³

*(1 mark)*

**b** Which of the processes shown in the diagram helps the body cool on a hot day?

.................................................................................................

.................................................................................................

*(1 mark)*

**c** Name another method the skin uses to cool the body.

.................................................................................................

*(1 mark)*

**(Total marks: 3)**

**2** Below are the results of a blood sugar test for two students. It records the one and a half hours following their mid-morning break in school.

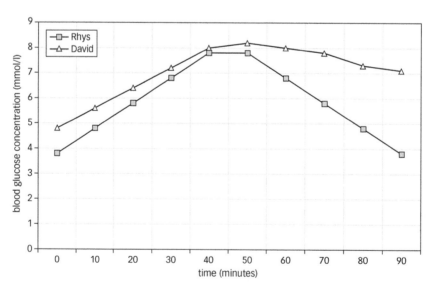

**a** What might have caused the blood sugar level to have risen in the blood of the two boys?

.................................................................................................

*(1 mark)*

**b** Which of the two boys do you think might be a diabetic? Explain why you think this.

Name of boy .......................................

Explanation ...........................................................................

............................................................................................................................................................

............................................................................................................................................................

*(3 marks)*

**c** In a healthy person:

**i** What is the name of the hormone that causes the blood sugar level to return to normal?

............................................................................................................................................................

**ii** What would happen to the amount of the hormone in a healthy person 30 minutes after a meal?

............................................................................................................................................................

**iii** Which organ of the body produces the hormone?

............................................................................................................................................................

**iv** Which organ of the body stores the sugar?

............................................................................................................................................................

**v** What is the sugar stored as?

............................................................................................................................................................

*(5 marks)*

**d** What causes the failure of control of blood sugar in a diabetic person?

............................................................................................................................................................

............................................................................................................................................................

............................................................................................................................................................

*(1 mark)*
***(Total marks: 10)***

**3** The kidney produces urine.

**a** What is the major waste product contained in urine?

..................................................................................................................................................

*(1 mark)*

**b** Describe how the production of urine by the kidneys can help to regulate the water concentration of the blood.

..................................................................................................................................................

..................................................................................................................................................

..................................................................................................................................................

..................................................................................................................................................

..................................................................................................................................................

*(4 marks)*

**c** When the kidneys fail, the patient needs treatment. There are two common treatments. Patients can have regular dialysis or they could have a kidney transplant.

The table below shows the concentrations of dissolved substances (measured in millimoles per litre, mmol/l) in the blood of a patient prior to dialysis and in the dialysis fluid being used during dialysis.

| Substance | Concentration in mmol/l | |
|---|---|---|
| | **Blood of kidney patient prior to dialysis** | **Kidney dialysis fluid** |
| Glucose | 5 | 5 |
| Urea | 25 | 0 |
| Sodium ions | 150 | 150 |
| Chloride ions | 150 | 150 |
| Potassium ions | 7 | 0 |

**i** The level of urea in a healthy person's blood is 5 mmol/l. What is the effect of this elevated value on a patient with kidney failure?

..................................................................................................................................................

*(1 mark)*

**ii** Suggest what will happen to the concentrations of dissolved substances in the patient's blood during dialysis, and explain why.

..................................................................................................................................................

..................................................................................................................................................

..................................................................................................................................................

..................................................................................................................................................

..................................................................................................................................................

*(3 marks)*

**d**   Kidney transplants are often preferred by patients. However, there are problems of rejection. Explain what causes rejection of an organ transplant, and how the problem can be overcome.

.......................................................................................................................................................................................

.......................................................................................................................................................................................

.......................................................................................................................................................................................

.......................................................................................................................................................................................

.......................................................................................................................................................................................

.......................................................................................................................................................................................

.......................................................................................................................................................................................

.......................................................................................................................................................................................

.......................................................................................................................................................................................

.......................................................................................................................................................................................

.......................................................................................................................................................................................

.......................................................................................................................................................................................

.......................................................................................................................................................................................

.......................................................................................................................................................................................

*(6 marks)*

***(Total marks: 15)***

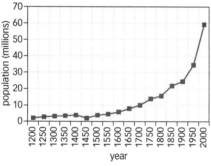

▲ The human population of the UK has been rising fast.

# The human population

In the UK, as in most of the world, the human **population** has shown a massive increase in the past few hundred years. This rapid increase is due to:

- improved diet
- improved hygiene
- improved healthcare
- a reduction in the infant mortality rate.

Graphs of the human population of the UK show that the population has more than doubled in the past 100 years.

## Sustainable living

As the population has increased in the past 100 years, the standard of living has also increased. There is a greater demand for manufactured products. This has placed a heavier demand on raw materials. The consequences of this are:

- the creation of more waste
- more land is used for building, farming, and extracting raw materials
- less fertile land is available for food production
- the world's resources are being used up faster than they can be replaced.

This lifestyle is now considered unsustainable.

Society has now recognised that they need to develop a more **sustainable** lifestyle. Sustainability is the use of resources without harming the environment by:

- replacing resources where possible, for example, replanting trees
- avoiding overuse of resources, for example, fishing quotas
- handling waste correctly to avoid **pollution**, such as recycling materials.

# Human impact on the environment

Human activity has two major impacts. It reduces the land available, and it releases pollutants into the environment. Most of these impacts come from either agriculture or the development of towns and industries.

## Loss of habitat

Humans need land for many reasons:

- building towns
- quarrying
- creating landfill waste dumps
- building industrial areas
- farming

In doing these things, they destroy the natural environment. This usually means cutting down natural woodland. The woodland provides food and shelter to many species. Without these habitats there will be a reduction in biodiversity (number and types of organisms).

## Pollution of land, air, and water

Pollutants are released into the environment. Different pollutants have different sources and different effects. Pollutants are released into the air, land, and water. The effects of many land pollutants occur when they are washed from the land into waterways.

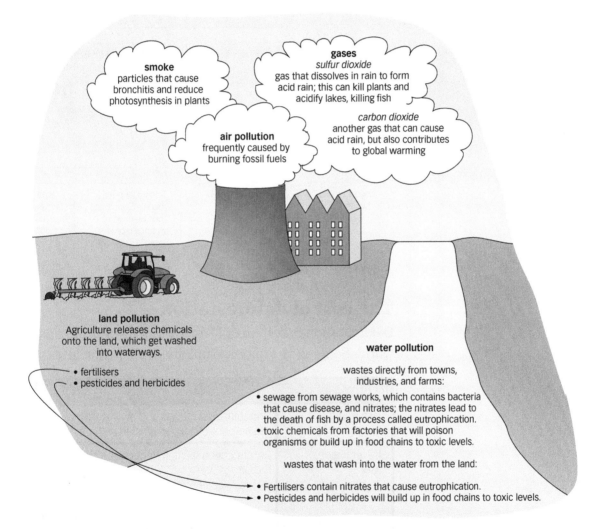

**smoke**
particles that cause bronchitis and reduce photosynthesis in plants

**gases**
*sulfur dioxide*
gas that dissolves in rain to form acid rain; this can kill plants and acidify lakes, killing fish

*carbon dioxide*
another gas that can cause acid rain, but also contributes to global warming

**air pollution**
frequently caused by burning fossil fuels

**land pollution**
Agriculture releases chemicals onto the land, which get washed into waterways.
• fertilisers
• pesticides and herbicides

**water pollution**

wastes directly from towns, industries, and farms:
• sewage from sewage works, which contains bacteria that cause disease, and nitrates; the nitrates lead to the death of fish by a process called eutrophication.
• toxic chemicals from factories that will poison organisms or build up in food chains to toxic levels.

wastes that wash into the water from the land:
• Fertilisers contain nitrates that cause eutrophication.
• Pesticides and herbicides will build up in food chains to toxic levels.

## Questions

1   What is the link between the human population size and the levels of pollution?

2   What are the main pollutants in the air?

3   **H** What is the effect of habitat destruction?

**Exam tip** AQA

For each pollutant you need to know the *chemical*, the *source*, and the *effect* on the environment.

## Revision objectives

- ✔ understand what deforestation is
- ✔ know the consequences of deforestation
- ✔ analyse and interpret environmental data

## Student book references

**3.25** Deforestation

## Specification key

✔ B3.4.2

# What is deforestation?

The large-scale felling of trees is called **deforestation**. This is happening worldwide, but is particularly common in countries like Brazil with the felling of tropical rainforests.

# Why does deforestation occur?

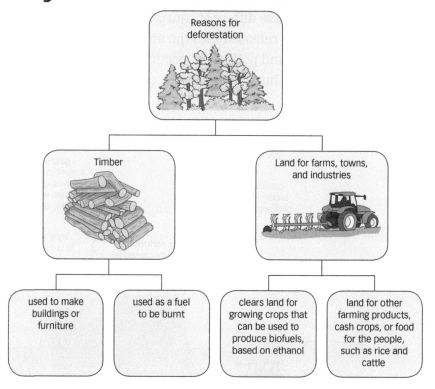

# The cost of deforestation

Deforestation is responsible for a number of problems in the world. Some of the major problems are listed below.

| Issue | Consequence |
|---|---|
| Slash and burn | Chopping down trees and burning the waste increases the amount of carbon dioxide in the atmosphere. |
| Effect on global gases | Deforestation leads to a rise in carbon dioxide levels in the atmosphere by:<br>• the release of carbon dioxide during burning<br>• the release of carbon dioxide during the decomposition of felled trees by microorganisms<br>• a reduction in photosynthesis, so less carbon dioxide is taken up by trees and locked up in wood for many years. |
| Reduction in biodiversity | A diverse forest community is removed and replaced with a single crop. This provides few habitats and removes large amounts of the same mineral from the soil. |
| Build-up of methane | Cattle farms and rice fields both release methane into the atmosphere. This contributes to global warming. |

## Destruction of peat bogs

A second natural habitat that is being destroyed by humans is the **peat** bogs.

- Peat is produced over thousands of years by the preservation of moss in wet boggy areas that become acidic.
- One major use of peat was in the production of nutrient-rich composts for gardeners and plant growers.
- Unfortunately the extracted peat dries, decays, and releases carbon dioxide into the atmosphere.
- Many gardeners now use peat-free compost.

## Analysing and interpreting environmental data

Frequently you will be asked to look at data on environmental topics and extract information from the data, or draw conclusions. Here is a simple example of this skill.

Below is a table of data about the extent and change of forest cover for a few countries.

| Country | Forest area in 2005 (1000 ha) | Annual change in area 2000 (1000 ha) | Percentage (%) change in forest area in 2000 | Annual change in area 2005 (1000 ha) | Percentage (%) change in forest area in 2005 |
|---|---|---|---|---|---|
| Brazil | 447 698 | −2681 | −0.62 | −3103 | −0.63 |
| Cambodia | 10 447 | −140 | −1.09 | −219 | −1.90 |
| UK | 2845 | +18 | +0.69 | +10 | +0.36 |
| Norway | 9387 | +17 | +0.19 | +17 | +0.18 |

▲ Data derived from forest resources assessment, 2005 (UN figures).

When you analyse the data, the figures indicate the following:

- Brazil and Cambodia are both experiencing deforestation – shown by the negative change in area.
- The UK and Norway are experiencing some reforestation – shown by the positive increase in area.
- The largest area of land being deforested is in Brazil – shown by the large value for the annual change in area.
- Cambodia has the quickest rate of loss – shown by the percentage-change figures.

Interpreting the meaning of this data suggests the following:

- Deforestation is a big issue in the tropical-rainforest areas of Brazil, as seen in the figures. This will result in the reduction of biodiversity and changes in global gas levels.
- This will have a greater effect in Brazil, because the area affected is greater.
- Reforestation is occurring in some countries, and this could be used as an example to other countries.

# Global warming

Global warming is the overall increase in average global temperatures. Most scientists think that the increase in global temperatures is caused by heat being trapped in the Earth's atmosphere by a layer of gases called greenhouse gases.

Human activity is greatly increasing the amounts of these gases in the atmosphere. Scientists are worried that this is increasing the rate of global warming. The two major gases are:

- carbon dioxide – released from burning fossil fuels, or from deforestation
- methane – from cattle, rice fields, and decaying waste.

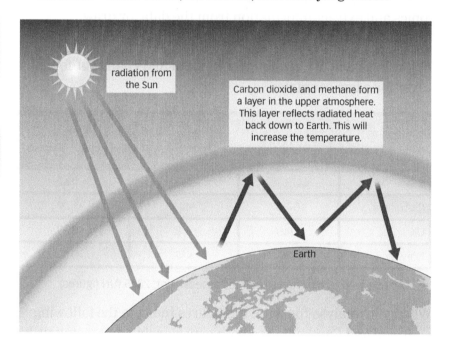

## The effects of global warming

A rise in global temperature of only a few degrees can have a major impact on the Earth.

## The effects of the oceans

Large bodies of water absorb large amounts of carbon dioxide. So oceans, lakes, and ponds remove carbon dioxide from the atmosphere, thus reducing its levels in the air. There are two ways this can happen:

- Phytoplankton absorb carbon dioxide during photosynthesis.
- Carbon dioxide dissolves in the water.

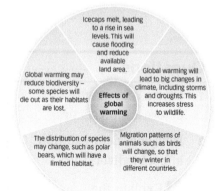

# Biofuels

Burning fossil fuels releases large amounts of carbon dioxide rapidly into the atmosphere, which has been locked up in the fuel for millions of years. Biofuels are a range of fuels from biological materials that are considered to be better for the environment. This is because the carbon dioxide produced when they burn is balanced by the carbon dioxide used in photosynthesis while the biological material is growing. So no overall increase in carbon dioxide in the atmosphere occurs. Biofuels include:

- wood
- alcohol
- biogas.

## Biogas production

Biogas is made by the anaerobic fermentation of the carbohydrates in plant material and sewage by bacteria. Biogas is effectively a fuel produced from human waste. Biogas is a mixture of gases:

- methane (50–75%)
- carbon dioxide (25–50%)
- hydrogen, nitrogen, and hydrogen sulfide (less than 10%).

Biogas production can vary in scale.

### Small-scale biogas production

In remote regions in third-world countries, families may have a small biogas digester to supply small amounts of fuel for cooking.

### Large-scale biogas production

Here large commercial tanks are used. Again there is an anaerobic fermentation of waste, but the waste is constantly added. This method has widespread use, including at sewage works in the UK. The rate of gas production is affected by climatic conditions, mainly temperature, working best at 32–35 °C. Large volumes of the gas are produced, which are used to power vehicles, generate electricity, and heat homes.

| Advantages of biofuels | Disadvantages of biofuels |
|---|---|
| Reduced fossil fuel consumption, by providing an alternative. | Causes habitat loss because large areas of land are needed to grow the plants. |
| No overall increase in levels of greenhouse gases, as the plants take in carbon dioxide to grow, and release it when burnt. | Habitat loss can lead to extinction of species. |
| Burning biogas and alcohol produces no particulates (smoke). | |

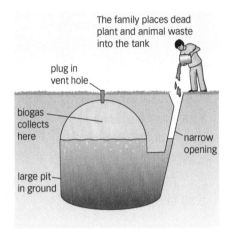

▲ A section through a biogas digester.

## Questions

1   Name a biofuel.

2   How does the burning of fossil fuels cause global warming?

3   **H** Discuss the effects of global warming on wildlife.

## Revision objectives

- ✓ know that fungi are used to produce foods such as mycoprotein
- ✓ understand the methods of maximising efficiency of production of farmed foods
- ✓ know how fishing has become a sustainable method of food production

## Specification key

✓ B3.4.4

## Exam tip

For each of the processes listed on these pages make sure you can evaluate the positive and negative effects in terms of food production for an increasing population.

# Feeding the multitudes

As the human population has increased new methods have been sought to produce enough food to feed everyone. The food needs to be produced as locally to the population as possible. This will reduce transport costs and pollution.

# Microbes as food

Fungi have always been used as a food source. By the 1960s scientists were becoming worried about the huge increase in the human population. If this increase continued, a consequence would be that there would not be enough protein to feed such a large population.

In 1967 a fungus was discovered called *Fusarium venenatum*, which could make a protein-based product. This protein is now called **mycoprotein**. This is produced in large **fermenters**.

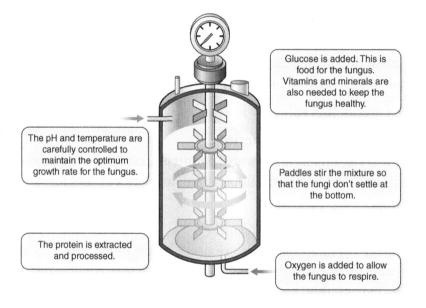

Glucose is added. This is food for the fungus. Vitamins and minerals are also needed to keep the fungus healthy.

The pH and temperature are carefully controlled to maintain the optimum growth rate for the fungus.

Paddles stir the mixture so that the fungi don't settle at the bottom.

The protein is extracted and processed.

Oxygen is added to allow the fungus to respire.

After production the protein is purified and made into many meat-substitute products.

# Energy-efficient farming

Farming produces the food for the human food chain. The problem is that at every link in the food chain, energy and biomass are lost. This means that the longer the food chain, the more biomass and energy has been lost because of the greater number of links. So food from the end of a food chain is less energy efficient than food from earlier in the chain.

## Reducing energy loss

Biologists have been able to suggest a number of methods that can reduce energy loss in farming:

- eating vegetable products, for example, flour, rather than eating animals fed on vegetable products – this removes a link in the chain
- intensive **factory-farming** methods such as battery farming
  - > reduces animal movement, reducing energy loss
  - > keeps animals warm by rearing them indoors and in large numbers, so less energy is lost to the surroundings.

# Fishing

Another common food is fish. As the population has increased, more fish have been caught from the world's oceans. Modern fishing fleets are very efficient at catching fish, using technology like sonar, and efficient, sophisticated nets. The impact of this is that the **fish stocks** in the oceans are in decline.

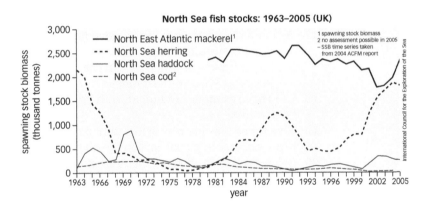

**North Sea fish stocks: 1963–2005 (UK)**

## Protecting fish stocks

Governments became concerned about these low fish stocks. Numbers were so low that the species would soon be unavailable to be caught as food. The fish populations needed to be maintained at a sufficient size so that breeding continued successfully. The two main conservation methods employed were:

- net size – increasing the hole size allowed younger fish through; these could survive and breed
- fishing **quotas** – governments limited the numbers of fish that could be caught; this maintained a breeding population.

These measures continue and have allowed fishing to become an example of sustainable food production.

## Advantages and disadvantages of factory-farming of animals

| Advantages | Disadvantages |
|---|---|
| Less energy is lost in the food chain, so more is available for human consumption. | Greater risk of disease spreading through the animals as they are in close contact. |
| Less labour intensive, as animals are all contained in a limited area. | Some people feel that the technique is inhumane, or cruel to the animals. |
| Less risk of attack from predators like foxes. | Some people believe that the quality of the product is poorer. |
| Production costs are cheaper. | |

**Key words**

mycoprotein, *Fusarium*, fermenter, factory farming, fish stocks, quotas

**Questions**

1  What is mycoprotein, and what is it used for?

2  Discuss how efficient fishing techniques have caused problems.

## Working to Grade E

1 Define population.
2 Below is a graph of the human population of the UK.

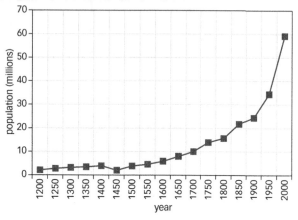

a What was the human population of the UK in 1700?
b How long did it take for the population of 1700 to double?
3 What is deforestation?
4 Give **two** uses of timber.
5 Peat bogs are habitats that are being destroyed by humans.
   a What do we use peat for?
   b Apart from loss of habitat, what other problem is created by the use of peat?
   c How can we overcome the problem?
6 What is global warming?
7 What is a biofuel?
8 Global warming is caused by greenhouse gases. What are the **two** main greenhouse gases?
9 What is the main constituent of biogas?
10 Below is a drawing of a small-scale biogas generator.

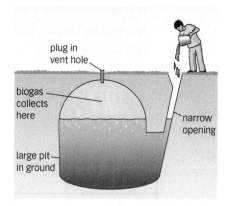

a What might people put into the generator to make the biogas?
b What do they mainly use the biogas for?
11 Global warming will have many effects, including affecting biodiversity. List **three** other effects of global warming.
12 What is a fermenter?
13 What is a fish quota?

## Working to Grade C

14 What are the **two** major ways in which humans affect the environment?
15 Look at the graph in question 2.
   a Explain why there was little increase in the population of the UK between 1200 and 1400.
   b The population of the UK is increasing.
      i During which century is the rate of increase the greatest?
      ii Suggest **three** reasons that might explain the increase during that century.
16 Complete the table below, which shows some of the major pollutants humans release into the air.

| Pollutant | Source | Effect on the environment |
|---|---|---|
| smoke | released from burning fossil fuels | |
| | | contributes to global warming, and acid rain |
| sulfur dioxide | | |

17 We are being encouraged to live sustainably.
   a What is sustainability?
   b Suggest ways in which we can change our lifestyle to become sustainable.
18 List **three** examples of human activity that cause habitat loss.
19 Give **three** reasons why deforestation contributes to an increase in global carbon dioxide levels.
20 Below is a graph showing the rates of deforestation of primary forests (natural forest) in various countries.

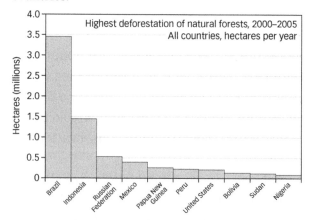

a The current annual rate of primary forest destruction in Brazil is 0.8% of the total forest. At this current rate, how long will it take for Brazil to completely lose its primary forests?
b Suggest reasons why Brazil might have undertaken such a high rate of deforestation.

21 Global warming is caused by greenhouse gases. What are the sources of the **two main** greenhouse gases?

22 Look back at the diagram in question 10.
 a What causes the biogas to be made inside the generator?
 b Why is this method of energy production particularly useful in remote parts of the world?

23 Global warming will have many effects. Explain how global warming will affect biodiversity.

24 What food source does the fungus *Fusarium* use in a fermenter?

25 Factory farming of animals is a common practice, for example, battery-farmed hens. Explain how battery farming of hens can reduce energy loss.

free range

battery

26 What has been the effect of increasing populations of humans on fish stocks?

27 Why is it more energy efficient to eat a producer than a carnivore?

28 Name **two** conditions that are controlled in a fermenter.

29 Give **two** methods by which trawler fleets became efficient.

30 Explain why scientists needed to find new sources of protein.

## Working to Grade A*

31 Habitats are being lost as a result of human activity. Explain why loss of habitat is of such concern to biologists.

32 Farmers use a number of chemicals on the land that wash into the waterways.
 a Describe the effect of fertilisers in the water.
 b Explain how small amounts of pesticides washed into the water can become a problem.

33 Explain why it is important to treat sewage at sewage works before it is released into our streams and rivers.

34 Look back at the graph in question 20.
 a Explain the relationship between deforestation and biodiversity.
 b The 7th greatest loss of primary forest in the world is seen in the USA. This is a loss of 0.2% of the primary forests. However, they record a total gain in forests of 0.1% per year. Suggest a reason for these figures.

35 Look back at the diagram in question 10. Give **two** significant differences between a small-scale and a large-scale biogas generator.

36 Explain how the oceans help to reduce carbon dioxide levels in the atmosphere.

37 Explain why using biofuels is believed not to contribute to global warming.

38 Look back at the diagram in question 25. What is the difference between battery farming and free-range farming?

39 How has increasing the hole size in nets allowed fish stocks to recover?

40 The distance involved in transporting foods before they are consumed is called 'food miles'. Explain why scientists are so concerned about this issue.

41 Evaluate the pros and cons of factory farming.

# Humans and their environment

**1** Humans carry out a number of processes that have a harmful effect on the environment.

List A gives a list of such processes.

List B gives the effects of these processes on the environment.

Draw one line from each process in list A to the effect in list B.

**List A**
**Process**

**List B**
**Effect on the environment**

| List A – Process | List B – Effect on the environment |
|---|---|
| Burning fossil fuels releasing sulfur dioxide. | Contributes to global warming. |
| Pesticides are used by farmers to kill pests. | Particles blacken leaves and reduce photosynthesis. |
| Release of sewage. | Dissolves in rain to form acid rain. |
| Growing large areas of rice that releases methane. | Builds up in the food chain, killing other organisms. |
| | Causes the death of fish by eutrophication. |

*(4 marks)*
***(Total marks: 4)***

**2** Some intensive farmers use a technique called battery or factory farming of animals like chickens to give improved yields.

Read the following points about factory farming.

- Battery farming of chickens is where large numbers of hens are reared in cages.
- This reduces the area of land needed, and fewer staff are employed to look after the birds.
- The chickens are kept disease-free by treatment with antibiotics.
- The birds are kept safe from predators because the cages are all kept indoors, in dark barns, using electric lights instead of daylight.
- To protect the birds from scratching each other in the small cages they often have their claws removed.

**a** Explain how the techniques used in factory farming of hens are a more energy-efficient way of farming.

.........................................................................................................................................................................

.........................................................................................................................................................................

.........................................................................................................................................................................

.........................................................................................................................................................................

.........................................................................................................................................................................

*(2 marks)*

**b**   Some scientists have concerns about the use of battery or factory farming. Suggest reasons why.

.......................................................................................................................................................

.......................................................................................................................................................

.......................................................................................................................................................

.......................................................................................................................................................

.......................................................................................................................................................

.......................................................................................................................................................

.......................................................................................................................................................

*(3 marks)*

*(Total marks: 5)*

**3**  Biogas is an example of a biofuel. The drawing shows a biogas generator used in remote towns in countries like Nepal.

**a**   Why is it important that air does not enter the generator?

....................................................................................

....................................................................................

....................................................................................

....................................................................................

*(1 mark)*

plug in
vent hole

biogas
collects
here

narrow
opening

large pit
in ground

**b**   Why is it important that the generator is insulated in the ground?

.......................................................................................................................................................

.......................................................................................................................................................

.......................................................................................................................................................

.......................................................................................................................................................

*(1 mark)*

**c**   What are the advantages of using this type of generator in a rural community?

.......................................................................................................................................................

.......................................................................................................................................................

.......................................................................................................................................................

.......................................................................................................................................................

.......................................................................................................................................................

*(2 marks)*

*(Total marks: 4)*

# Ethical issues raised by scientific developments

Research scientists are frequently developing theories and practices that have a direct impact on people's lives. For example, in medical research treatments are developed that can dramatically improve a patient's quality of life. On the face of it, medical developments like this seem to be for the good, and there would be no reason to think more about it. However, many medical developments can create a number of ethical, economic, or social issues. It may be the role of a clinical scientist to make decisions based on evidence to maximise the developments and save lives. Such an example can be seen in the treatments used for kidney failure.

# Treating kidney failure

Patients with kidney failure have two major forms of treatment open to them:

- Kidney dialysis – This technique was first developed in 1943 by Dr Koff in the Netherlands during World War II. Since this time many improvements have been made to the technique. The process saves lives but requires several treatments a week, each for a few hours. However, this treatment does not provide a cure for chronic kidney failure.
- Kidney transplant – In 1954 Dr Joseph Murray performed a new treatment in Boston, US, where kidneys were transplanted from one individual to another. In general this treatment is superior, as it doesn't require regular, time-consuming dialysis sessions. People with transplants can lead normal lives post surgery.

It would seem that transplants are the perfect solution. However, there are a number of issues raised by the treatment.

## The implications of kidney transplants

### Cost

Transplant surgery is very expensive. In 2009 the typical cost was £17000 per patient, with £5000 drug care per year. However, in the long term the surgical option might work out cheaper than dialysis, which costs £30000 per year.

### Number of donors

In 2009 there were 2497 kidney transplants in the UK. However, there were 6920 people still waiting for transplant. The need for kidneys outstrips availability.

### Organ trade

In some countries people can sell their organs. However, the illegal sale of organs is an ethical issue in the UK.

### Rejection

Tissue typing is needed in organ transplants. The closer the match, the more successful the anti-rejection (immunosuppressant) therapy.

### Patient's state of health

Clinicians must evaluate a patient's state of health and age, to assess the likely success, and improvement to quality of life, of organ transplant. Some people raise concerns about transplanting into people with histories of long-term drug abuse.

### Religious objections

Some religious beliefs prevent some patients having any form of transplant. Personal beliefs must be respected.

## Reaching a decision

So, who gets the organ?

A clinical doctor must make a decision quickly. To do this they scrutinise the evidence available for each case.

Guidelines have been created by groups of doctors. These guidelines allow judgements to be made by individual transplant surgeons to decide which patient is best suited for a transplant. The decision uses social and scientific issues to select a patient who will gain the most, and thus there is less risk of wasting limited organs. The guidelines use scientific evidence and so prevent public or political disquiet.

# AQA Upgrade

## Dissecting exam questions

The skin plays a role in temperature control in the body. Look at the diagram of the skin.

1 Explain how sweating can regulate body temperature when the body overheats and overcools. *(3 marks)*

2 Explain **one** other way in which the structure of the skin is involved in temperature regulation. *(2 marks)*

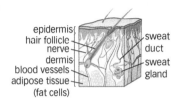

epidermis
hair follicle
nerve
dermis
blood vessels
adipose tissue
(fat cells)
sweat duct
sweat gland

**QUESTION**

---

**G–E**

1 We sweat in the heat. This means we sweat more when its hot. This helps us in the heat.
2 When we get cold we shiver. This helps to warm us up because we are cold.

**Examiner:** This student has only told us one thing: that we sweat more in the heat. They have repeated the same point. The candidate has not dissected the question. They missed the command word, and have not explained the effect of sweating. They have also not seen the two sections – overheating and overcooling. 1 mark would be awarded.

This student has missed the instruction to identify a structure in the skin that helps temperature control. They did not need to have much additional knowledge as the diagram in the question lists structures. They have instead opted to talk about muscles. Therefore they gain no marks.

---

**D–C**

1 In the heat we produce lots more sweat. This is because it covers the skin and takes heat from the skin and evaporates. This cools our bodies. When we sweat on hot days this will make us thirsty.
2 The skin also has blood vessels that help to control the temperature.

**Examiner:** This student has picked up on the command word, and explained the effect of sweating. This has been done too much. The candidate has set out to say all they know about sweating. However, the candidate has missed the final section of the question, and not mentioned overcooling. The mark scheme will divide the marks up allotting some marks for both sections. Therefore, having missed the overcooling, they will gain only 2 marks.

This candidate has identified a structure in the skin that is involved in temperature regulation. But this time they have missed the command word to explain. Only 1 mark awarded.

---

**B–A\***

1 When we overheat, we will sweat more. This will cool our body down, by evaporating off the skin. But when we overcool, little or no sweat will be produced because we do not need to cool down.
2 There are blood vessels in the skin which are involved in temperature control. They can dilate when it is hot to lose heat, or constrict when it is cold to retain heat.

**Examiner:** This is a clear and logical answer. This question has three sections in it. First the command word 'explain' means tell the examiner how something happens. Secondly, the question asks about overheating and finally, overcooling. The candidate has dealt with both parts of the question, overheating and overcooling. They have noted the key word 'explain'. 3 marks awarded.

This is a well-structured answer. The question can again be dissected into parts. The command word 'explain' expects a statement of how the temperature is controlled. The next part of the question instructs the candidate to identify a structure in the skin involved in temperature control. Finally, the candidate should link that structure to its role in temperature regulation. This has been achieved here. 2 marks awarded.

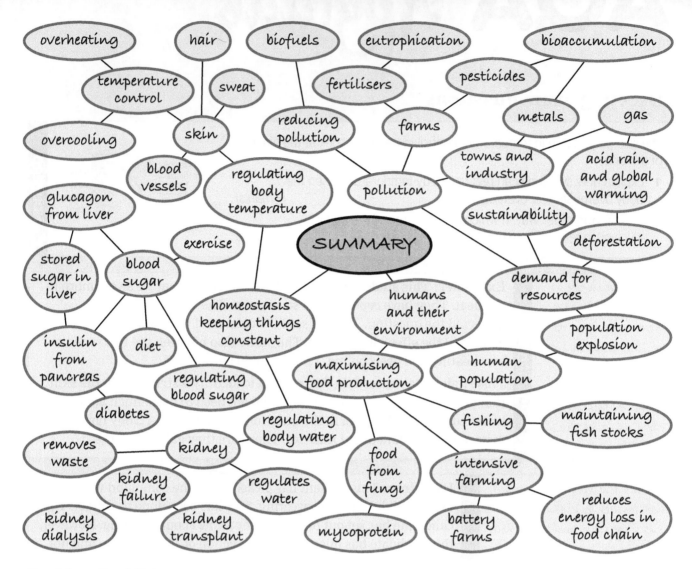

## Revision checklist

○ Animals need to regulate their internal environment by getting rid of wastes like urea and carbon dioxide, and maintaining constant internal conditions.

○ Homeostasis is the process where animals maintain constant internal levels for body temperature, water, ions, and sugar.

○ The kidney is an organ that filters wastes like urea from the blood and regulates the levels of water and ions in the blood.

○ Kidney failure leads to a build-up of toxic wastes like urea, which can be fatal.

○ Kidney failure can be treated by dialysis, where the blood is purified using a dialysis machine.

○ Kidney transplants replace the damaged kidney but the transplanted kidney must be matched by tissue typing, to avoid rejection.

○ The skin is the organ involved in temperature control.

○ Overheating is prevented by sweating, hairs lying flat, and vasodilation.

○ Overcooling is prevented by shivering, insulation, and vasoconstriction.

○ Blood glucose levels are controlled by the hormones insulin and glucagon.

○ Diabetes is a condition where insulin is not made, resulting in a failure to control the blood glucose level.

○ The human population is increasing, which is putting extra demand on the Earth's resources.

○ Humans release pollutants into the air and water from agriculture, towns, and industry.

○ Deforestation is the process of removing large areas of natural woodland. This can result in an increase in carbon dioxide levels in the atmosphere, and loss of habitat.

○ Carbon dioxide, and methane from rice fields and cattle, are greenhouse gases that lead to global warming.

○ Biofuels are fuels from biological materials, which do not contribute to global warming.

○ Humans are increasing food production to feed the increasing population by producing protein foods from fungi like mycoprotein.

○ Farming methods can be made energy efficient by reducing energy losses in food chains.

○ Overfishing had led to dwindling fish stocks. Governments introduced quotas and changes to nets to prevent overfishing.

## Newlands octaves

In the 1860s, John Newlands listed the 56 elements then known in order of atomic weight. Every eighth element had similar properties. He used this pattern to group the elements. Newlands called his discovery the **law of octaves**.

Other chemists criticised the law of octaves. They said that some groups included an element whose properties were very different to the other elements in the group. For example, nickel was grouped with fluorine, chlorine, and bromine.

## Mendeleev's periodic table

In 1869 Dmitri Mendeleev created the first **periodic table**. He arranged the elements in order of increasing atomic weight, and grouped elements with similar properties together. But some elements seemed to be in the wrong groups. He overcame this problem by:

* swapping the positions of some elements, for example, iodine and tellurium
* leaving gaps for elements that he predicted did exist, but that had not been discovered.

Later, other scientists discovered some of the missing elements. This increased scientists' confidence that the periodic table was a useful tool.

## The modern periodic table

Early in the twentieth century, scientists discovered the particles that make up atoms – protons, neutrons, and electrons.

| 1 | 2 | | | | | | | | | | | 3 | 4 | 5 | 6 | 7 | 0 |
|---|---|---|---|---|---|---|---|---|---|---|---|---|---|---|---|---|---|
| | | | | | | | 1 **H** Hydrogen 1 | | | | | | | | | | 4 **He** Helium 2 |
| 7 **Li** Lithium 3 | 9 **Be** Beryllium 4 | | | | | | | | | | | 11 **B** Boron 5 | 12 **C** Carbon 6 | 14 **N** Nitrogen 7 | 16 **O** Oxygen 8 | 19 **F** Fluorine 9 | 20 **Ne** Neon 10 |
| 23 **Na** Sodium 11 | 24 **Mg** Magnesium 12 | | | | | | | | | | | 27 **Al** Aluminium 13 | 28 **Si** Silicon 14 | 31 **P** Phosphorus 15 | 32 **S** Sulfur 16 | 35.5 **Cl** Chlorine 17 | 40 **Ar** Argon 18 |
| 39 **K** Potassium 19 | 40 **Ca** Calcium 20 | 45 **Sc** Scandium 21 | 48 **Ti** Titanium 22 | 51 **V** Vanadium 23 | 52 **Cr** Chromium 24 | 55 **Mn** Manganese 25 | 56 **Fe** Iron 26 | 59 **Co** Cobalt 27 | 59 **Ni** Nickel 28 | 63.5 **Cu** Copper 29 | 65 **Zn** Zinc 30 | 70 **Ga** Gallium 31 | 73 **Ge** Germanium 32 | 75 **As** Arsenic 33 | 79 **Se** Selenium 34 | 80 **Br** Bromine 35 | 84 **Kr** Krypton 36 |
| 85 **Rb** Rubidium 37 | 88 **Sr** Strontium 38 | 89 **Y** Yttrium 39 | 91 **Zr** Zirconium 40 | 93 **Nb** Niobium 41 | 96 **Mo** Molybdenum 42 | [98] **Tc** Technetium 43 | 101 **Ru** Ruthenium 44 | 103 **Rh** Rhodium 45 | 106 **Pd** Palladium 46 | 108 **Ag** Silver 47 | 112 **Cd** Cadmium 48 | 115 **In** Indium 49 | 119 **Sn** Tin 50 | 122 **Sb** Antimony 51 | 128 **Te** Tellurium 52 | 127 **I** Iodine 53 | 131 **Xe** Xenon 54 |
| 133 **Cs** Caesium 55 | 137 **Ba** Barium 56 | 139 **La*** Lanthanum 57 | 178 **Hf** Hafnium 72 | 181 **Ta** Tantalum 73 | 184 **W** Tungsten 74 | 186 **Re** Rhenium 75 | 190 **Os** Osmium 76 | 192 **Ir** Iridium 77 | 195 **Pt** Platinum 78 | 197 **Au** Gold 79 | 201 **Hg** Mercury 80 | 204 **Tl** Thallium 81 | 207 **Pb** Lead 82 | 209 **Bi** Bismuth 83 | [209] **Po** Polonium 84 | [210] **At** Astatine 85 | [222] **Rn** Radon 86 |
| [223] **Fr** Francium 87 | [226] **Ra** Radium 88 | [227] **Ac*** Actinium 89 | [261] **Rf** Rutherfordium 104 | [262] **Db** Dubnium 105 | [266] **Sg** Seaborgium 106 | [264] **Bh** Bohrium 107 | [277] **Hs** Hassium 108 | [268] **Mt** Meitnerium 109 | [271] **Ds** Darmstadtium 110 | [272] **Rg** Roentgenium 111 | | | | | | | |

Key: relative atomic mass / **atomic symbol** / name / atomic (proton) number

Elements with atomic numbers 112–116 have been reported but not fully authenticated

\* The Lanthanides (atomic numbers 58–71) and the Actinides (atomic numbers 90–103) have been omitted.
**Cu** and **Cl** have not been rounded to the nearest whole number.

▲ The modern periodic table. It is called a periodic table because elements with similar properties occur at regular intervals. Elements with similar properties are in columns, called groups. The horizontal rows are called periods.

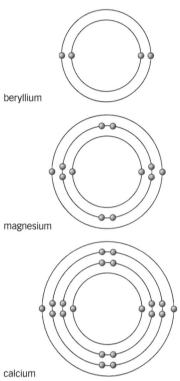

beryllium

magnesium

calcium

▲ The Group 2 elements all have two electrons in their highest occupied energy level.

The scientists arranged the elements in order of increasing **atomic number** (proton number) to make a new periodic table. All the elements were now in appropriate groups. The problems of Mendeleev's periodic table, based on atomic weights, had been solved.

## Electronic structure and the periodic table

An element's position in the periodic table is linked to its electronic structure. Elements in the same group have the same number of electrons in their highest occupied energy level (outer shell).

For the main groups, the number of electrons in the highest occupied energy level is equal to the group number.

## The transition elements

The **transition elements** form the central block of the periodic table.

the transition elements

| | | | | | | | | | | | | | | | | | |
|---|---|---|---|---|---|---|---|---|---|---|---|---|---|---|---|---|---|
| | | | | | | | | H | | | | | | | | | He |
| Li | Be | | | | | | | | | | | B | C | N | O | F | Ne |
| Na | Mg | | | | | | | | | | | Al | Si | P | S | Cl | Ar |
| K | Ca | Sc | Ti | V | Cr | Mn | Fe | Co | Ni | Cu | Zn | Ga | Ge | As | Se | Br | Kr |
| Rb | Sr | Y | Zr | Nb | Mo | Tc | Ru | Rh | Pd | Ag | Cd | In | Sn | Sb | Te | I | Xe |
| Cs | Ba | La | Hf | Ta | W | Re | Os | Ir | Pt | Au | Hg | Tl | Pb | Bi | Po | At | Rn |
| Fr | Ra | Ac | Rf | Db | Sg | Bh | Hs | Mt | Ds | Rg | | | | | | | |

The transition elements are metals. Most transition elements:
• are strong and hard
• have high densities
• have high melting points (except for mercury, which is liquid at room temperature).

The transition elements react slowly – if at all – with water and oxygen:
• Platinum and gold do not react with water and oxygen.
• Iron reacts slowly with water and oxygen at room temperature to make rust (hydrated iron(III) oxide).

Many transition elements form ions with different charges. For example, copper has two oxides:
• CuO is black. It includes a $Cu^{2+}$ ion.
• $Cu_2O$ is red. It includes a $Cu^+$ ion.

Many transition elements form coloured compounds. For example, vanadium compounds can be yellow, blue, green, or lilac.

Transition elements and their compounds are useful catalysts. For example, vanadium pentoxide ($V_2O_5$) speeds up a vital step in the manufacture of sulfuric acid.

## Physical properties

The Group 1 elements are called the **alkali metals.** The group includes lithium, sodium, and potassium.

Group 1
the alkali metals

| | | | | | | | | | | | | | | | | | H | | He |

▲ Group 1 is on the left of the periodic table.

The alkali metals are soft – you can cut them with a knife. They also have low densities. The densities of lithium, sodium, and potassium are so low that they can float on water.

The alkali metals have low melting and boiling points compared to all transition elements except mercury. The further down Group 1 an element is, the lower its melting and boiling points.

| Name of element | Melting point (°C) | Boiling point (°C) |
| --- | --- | --- |
| lithium | 180 | 1330 |
| sodium | 98 | 890 |
| potassium | 64 | 774 |
| rubidium | 39 | 688 |

## Reactions with non-metals

The alkali metals react with non-metals such as chlorine. For example, sodium burns vigorously in chlorine. The product is sodium chloride (common salt).

$$\text{sodium} + \text{chlorine} \rightarrow \text{sodium chloride}$$
$$2Na\,(s) + Cl_2\,(g) \rightarrow 2NaCl\,(s)$$

The alkali metals also react with oxygen, another non-metal.
- At room temperature, their surfaces tarnish when exposed to air. This is why they are stored in oil.
- On heating, they react vigorously with oxygen from the air. For example:

$$\text{sodium} + \text{oxygen} \rightarrow \text{sodium oxide}$$
$$4Na\,(s) + O_2\,(g) \rightarrow 2Na_2O\,(s)$$

### Revision objectives
- ✔ compare the properties of Group 1 elements and transition elements
- ✔ describe the reactions of the Group 1 elements with non-metals
- ✔ explain the trend in reactivity of the Group 1 elements

### Student book references
- **3.3** Alkali metals – 1
- **3.4** Alkali metals – 2

### Specification key
- ✔ C3.1.3 a – b and h (part)

# Alkali metal compounds

A compound made up of an alkali metal and a non-metal is **ionic**. The metal ion has a charge of +1. For example:

- Potassium chloride (KCl) is made up of potassium ions ($K^+$) and chloride ions ($Cl^-$). In a crystal of the compound, there is one potassium ion for every chloride ion.
- Sodium oxide ($Na_2O$) is made up of sodium ions ($Na^+$) and oxide ions ($O^{2-}$). In a crystal of the compound, there are two sodium ions for every one oxide ion.

The compounds are white solids at room temperature. They dissolve in water to form colourless solutions.

# Reactions with water

The alkali metals react vigorously with water. The products are:

- hydrogen gas
- a hydroxide.

For example:

$$sodium + water \rightarrow sodium\ hydroxide + hydrogen$$
$$2Na\ (s) + 2H_2O\ (l) \rightarrow 2NaOH\ (aq) + H_2\ (g)$$

Alkali metal hydroxides dissolve in water to give alkaline solutions.

# Group trend

The Group 1 elements are more reactive than the transition elements. So their reactions with oxygen and water are more vigorous than those of the transition elements.

The further down Group 1 an element is, the more vigorous its reactions.

> **H** This trend can be explained by the energy level of the outer electrons. Potassium is lower down Group 1 than lithium. The outermost electron of potassium is in a higher energy level than that of lithium. This means that, in reactions, potassium gives away its outermost electron more easily than lithium. Potassium is more reactive than lithium.

## Exam tip

The Group 1 elements are softer and weaker than the transition elements. They have lower densities and lower melting points. The Group 1 elements react more vigorously with water and oxygen than the transition metals.

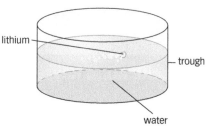

lithium — trough

water

▲ All the alkali metals react vigorously with water. The reactions get more vigorous going down the group. Lithium, at the top, zooms around on the water surface. Caesium, at the bottom, reacts explosively with water.

## Questions

1  Draw a table to show **three** differences in the properties of the Group 1 elements and the transition elements.

2  Describe the differences and similarities between the reaction of lithium with water and the reaction of potassium with water.

3  **H** The further down Group 1 an element is, the more vigorous its reactions. Explain why.

## Physical properties

The Group 7 elements are called the **halogens**. The group includes fluorine, chlorine, bromine, and iodine.

| | | | | | | | | | | | | | | | | Group 7 the halogens | |
|---|---|---|---|---|---|---|---|---|---|---|---|---|---|---|---|---|---|

|  |  |  |  |  |  |  |  |  |  |  |  |  |  |  |  | H |  |  |  |  |  |  | He |
|----|----|----|----|----|----|----|----|----|----|----|----|----|----|----|----|----|----|
| Li | Be |  |  |  |  |  |  |  |  |  |  | B | C | N | O | F | Ne |
| Na | Mg |  |  |  |  |  |  |  |  |  |  | Al | Si | P | S | Cl | Ar |
| K | Ca | Sc | Ti | V | Cr | Mn | Fe | Co | Ni | Cu | Zn | Ga | Ge | As | Se | Br | Kr |
| Rb | Sr | Y | Zr | Nb | Mo | Tc | Ru | Rh | Pd | Ag | Cd | In | Sn | Sb | Te | I | Xe |
| Cs | Ba | La | Hf | Ta | W | Re | Os | Ir | Pt | Au | Hg | Tl | Pb | Bi | Po | At | Rn |
| Fr | Ra | Ac | Rf | Db | Sg | Bh | Hs | Mt | Ds | Rg |  |  |  |  |  |  |  |

▲ Group 7 is towards the right of the periodic table.

The Group 7 elements exist as diatomic molecules, for example, $Cl_2$.

The further down the group an element is, the higher its melting point and boiling points.

| Name of element | Melting point (°C) | Boiling point (°C) | State at room temperature | Colour |
|---|---|---|---|---|
| chlorine | −101 | −34.7 | gas | green |
| bromine | −7.20 | 58.8 | liquid | orange/brown |
| iodine | 114 | 184 | solid | grey/black with violet vapour |

## Reactions with metals

The halogens react with metals to form ionic compounds. For example, chlorine reacts with iron to form iron chloride.

$$iron + chlorine \rightarrow iron\ chloride$$
$$2Fe\,(s) + 3Cl_2\,(g) \rightarrow 2FeCl_3\,(s)$$

### Questions

1 Describe the trends in the melting points and boiling points of the Group 7 elements.

2 Predict which of the reactions below is more vigorous. Give a reason for your decision.
iron + chlorine → iron chloride
iron + iodine → iron iodide

3 Write equations to summarise the displacement reactions of:
a chlorine with sodium iodide solution
b bromine with potassium iodide solution.

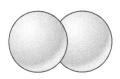

▲ Chlorine molecules are made up of two chlorine atoms joined together by a strong covalent bond.

Iron chloride is made up of two types of ion:
- iron ions, $Fe^{3+}$
- chloride ions, $Cl^-$.

The chloride ion is an example of a **halide ion**. Bromine forms bromide ions ($Br^-$) and iodine forms iodide ions ($I^-$). All halide ions have a charge of $-1$.

In Group 7, the further down the group an element is, the less reactive the element.

**H** This trend can be explained by the energy levels of the outer electrons in halogen atoms. When iron reacts with chlorine, iron atoms give each chlorine atom one extra electron. The electron completes the outer energy level of the chlorine atom, forming a chloride ion. The higher the energy level of the outer electrons, the less easily electrons are gained.

## Displacement reactions

In displacement reactions, a more reactive halogen displaces a less reactive halogen from an aqueous solution of its salt. For example, chlorine is more reactive than bromine, so:

chlorine + potassium bromide $\rightarrow$ potassium chloride + bromine

$Cl_2$ (aq) + 2KBr (aq) $\rightarrow$ 2KCl (aq) + $Br_2$ (aq)

# Questions
## The periodic table

### Working to Grade E

1   Highlight the correct word or phrase in each pair of **bold** words.
Copper is a **Group 1/transition** element. It forms ions with charges that are **different/the same.** Its compounds are **coloured/white.** Copper and its compounds make useful **alkalis/catalysts.**

2   Give the number of electrons in the highest occupied energy level (outer shell) of the elements in:
a   Group 1      b   Group 3      c   Group 7

3   Write T next to the statement that is true. Write corrected versions of the statements that are false.
a   In the periodic table, a vertical column is called a period.
b   In early periodic tables, the elements were arranged in order of atomic weight.
c   Newlands left gaps in his periodic table for elements he predicted did exist, but had not yet been discovered.

4   Use your knowledge of the properties of the Group 1 elements, and the transition elements, to answer the questions below. You will also need the periodic table. Draw a ring around:
a   the hardest metal in this list: sodium, manganese, lithium
b   the metal with the lowest melting point in this list: copper, potassium, gold, chromium
c   the metal with the highest density in this list: sodium, iron, lithium, potassium
d   the metal that reacts least vigorously with water in this list: manganese, potassium, caesium, rubidium.

### Working to Grade C

5   Complete the table below.

| Name of compound | Formula of metal ion in compound | Appearance and state of compound at room temperature |
|---|---|---|
| potassium chloride | | |
| sodium bromide | | |
| lithium chloride | | |

6   Complete the word equations.
a   sodium + chlorine → _____
b   lithium + oxygen → _____
c   sodium + water → _____ + _____
d   potassium + water → _____ + _____

7   Decide which of the following pairs of solutions will react in displacement reactions. Then write word equations for the pairs of solutions that react.
a   chlorine and potassium bromide
b   iodine and potassium bromide
c   bromine and potassium iodide
d   bromine and sodium chloride
e   chlorine and potassium iodide

8   For the sentences below, write **1** next to each sentence that is true for Group 1. Write **7** next to each sentence that is true for Group 7. To help you, use the data in the table, your own knowledge, and the periodic table.

| Element | Boiling point (°C) | Density (g/cm³) |
|---|---|---|
| lithium | 1330 | 0.53 |
| sodium | 890 | 0.97 |
| potassium | 774 | 0.86 |
| chlorine | −35 | 1.56 |
| bromine | 59 | 3.1 |
| iodine | 184 | 4.9 |

a   Going down this group, boiling point decreases.   ☐
b   Going down this group, the elements get less reactive   ☐
c   The elements in this group form ionic compounds with non-metals.   ☐
d   The elements in this group form ionic compounds in which the negative ion carries a charge of –1.   ☐
e   The top three elements of this group are less dense than water.   ☐
f   Going down this group, the elements get more reactive.   ☐

### Working to Grade A*

9   Write balanced symbol equations for the reactions shown by the word equations below.
a   sodium + bromine → sodium bromide
b   lithium + water → lithium hydroxide + hydrogen
c   chlorine + potassium iodide → potassium chloride + iodine

10   Alice has written a paragraph to explain the trend in the reactivities of the Group 7 elements. There is **one** mistake in **each** sentence. Write out the sentences, correcting the mistakes.
When a metal reacts with a halogen, the metal atoms give each halogen atom two extra electrons. This electron completes the inner energy level of the halogen atom. The closer the outer energy level is to the nucleus, the smaller the attraction between the newly added electrons and the nucleus. So the lower down Group 7 an element is, the more easily its atoms gain electrons, and the less reactive the element is.

**1** This question is about the periodic table.

**a** Read the information in the box below. Then answer the questions that follow.

> In 1860 an Italian scientist, Cannizzaro, published a list of the atomic weights of all the elements then known.
>
> In 1869, a Russian scientist, Mendeleev arranged the elements in order of atomic weight. He grouped together elements with similar properties.
>
> He left gaps for elements that he predicted did exist, but that had not been discovered.
>
> In 1875 a French scientist discovered an element to fill one of Mendeleev's gaps. He called it gallium. Four years later, a Swedish scientist discovered another of the missing elements. He called it scandium.

    **i** Explain how Mendeleev's work relied on the findings of Cannizzaro.

    ...................................................................................................................................................

    ...................................................................................................................................................

    *(1 mark)*

    **ii** Suggest why the discoveries of gallium and scandium in the 1870s made scientists more confident that the periodic table of the time was a useful tool.

    ...................................................................................................................................................

    ...................................................................................................................................................

    *(1 mark)*

**b**  **i** Write **two** words to complete the sentence below.

    In the modern periodic table, the elements are arranged in order of their ......................... .........................

    *(1 mark)*

    **ii** Explain how, in the modern periodic table, the position of an element in the periodic table is linked to its electronic structure.

    In your answer, include examples and give the electronic structure of at least one element.

    ...................................................................................................................................................

    ...................................................................................................................................................

    ...................................................................................................................................................

    *(3 marks)*
    **(Total marks: 6)**

**2** Describe some differences in the properties of the Group 1 elements and the transition elements.

Include at least **two** word equations, or balanced symbol equations, to illustrate your answer.

*In this question you will get marks for using good English, organising information clearly, and using scientific words correctly.*

...................................................................................................................................................

...................................................................................................................................................

...................................................................................................................................................

...................................................................................................................................................

...................................................................................................................................................

...................................................................................................................................................

...................................................................................................................................................

...................................................................................................................................................

...................................................................................................................................................

...................................................................................................................................................

*(6 marks)*

***(Total marks: 6)***

**3** The table gives some properties of the Group 7 elements.

| Element | Melting point (°C) | Boiling point (°C) | Reaction with hydrogen |
|---------|--------------------|--------------------|------------------------|
| fluorine | −220 | −118 | On mixing the two elements explode. Hydrogen fluoride is formed. |
| chlorine | −101 | −35 | A mixture of the two elements explodes when a camera flashes. Hydrogen chloride is formed. |
| bromine | −7 | 59 | A mixture of the two elements burns quickly when ignited by a lighted splint. |

**a** Use the data in the table to give the state of fluorine at room temperature (20 °C)

...................................................................................................................................................

*(1 mark)*

**b** Describe the trend in boiling point as the group is descended from top to bottom.

...................................................................................................................................................

*(1 mark)*

**c** Write a word equation for the reaction of hydrogen with fluorine.

Select information from the table to help you.

......................................................................................................................................................................

*(1 mark)*

**d** **i** Use the periodic table to identify the element below bromine in Group 7.

......................................................................................................................................................................

*(1 mark)*

**ii** Select data from the table above to predict how this element reacts with hydrogen.

Include the name of the product and suggest how vigorously the two elements will react.

......................................................................................................................................................................

......................................................................................................................................................................

*(2 marks)*

**e** **i** Describe the trend in reactivity in the Group 7 elements.

Illustrate your answer by referring to:
- reactions shown in the table at the start of question 3
- the displacement reactions of the Group 7 elements.

......................................................................................................................................................................

......................................................................................................................................................................

......................................................................................................................................................................

......................................................................................................................................................................

*(4 marks)*

**H** **ii** Explain the trend in reactivity you have described in part **i** above.

......................................................................................................................................................................

......................................................................................................................................................................

*(2 marks)*

***(Total marks: 12)***

# Making water safe to drink

Water of the correct quality is vital for life. Microbes and dissolved salts may harm human health.

In the UK, water companies provide safe drinking water by:

- *Choosing a suitable source:* Water from different sources needs different treatments. Water from boreholes may have been filtered by underground rock for many years. It would then need little treatment. Canal water may contain algae, bacteria, and fertiliser chemicals. It needs careful treatment.
- *Filtration:* Water trickles through huge beds of sand, called **filter beds**. These remove solid impurities.
- *Sterilisation:* Chlorine **sterilises** water by killing microbes. The chlorine remains in the water. It kills any microbes that get into the water between the treatment works and people's homes.

▲ The flow diagram summarises the stages by which water is made safe to drink.

# Adding fluorides

Some water companies add fluoride compounds to tap water. Fluorides help to prevent tooth decay.

There are arguments for and against adding fluoride compounds to water.

*Arguments for adding fluoride to drinking water*

- It prevents tooth decay, meaning fewer people suffer from toothache and other dental problems.
- Less money is spent treating dental problems.

*Arguments against adding fluoride to drinking water*

- It is expensive.
- If everyone looked after their teeth properly there would be little need to add fluorides to water.
- Swallowing very large amounts of fluoride compounds may make teeth go yellow.

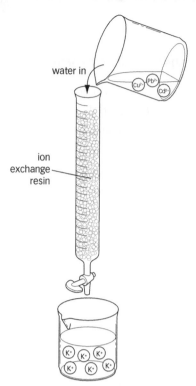

water in

ion exchange resin

In this water filter, metal ions are replaced by potassium ions. The ions are not drawn to scale.

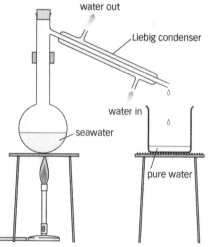

water out

Liebig condenser

water in

seawater

pure water

The diagram shows how to produce small amounts of pure water from seawater. Countries such as the United Arab Emirates, which has no rivers or lakes, obtain most of their water by the large-scale distillation of seawater.

## Key words

**filter bed, sterilises, ion exchange resin, activated carbon, adsorption, distillation**

# Home water filters

Some people use water filters to improve the taste or quality of their tap water.

## Ion exchange resin filters

Some types of water filter contain an **ion exchange resin**. Ion exchange resins remove metal ions (for example, lead, copper or cadmium) from the water. The filter replaces these metal ions with less harmful ions, such as hydrogen, sodium, or potassium.

Ion exchange resins are also used to soften hard water.

### Carbon filters

Carbon water filters remove chlorine from water, as well as other substances with unpleasant smells or tastes. The filters contain **activated carbon**. This has a huge surface area. As tap water passes through the filter, molecules of unwanted substances stick to the surface of the carbon. The process is called **adsorption**.

### Silver filters

Some home water filters contains a source of silver ions, $Ag^+$. These destroy some types of dangerous bacteria.

# Distillation

Pure water can be produced by the **distillation** of seawater. The process requires huge energy inputs. This makes it very expensive.

## Questions

1 List the **three** main stages by which water of the correct quality is provided for British homes.

2 Evaluate the advantages and disadvantages of adding fluoride to water.

3 Draw a table to summarise the substances removed from tap water by these types of water filter – carbon, silver, ion exchange resin.

## Measuring water hardness

**Soft water** easily forms lather with soap. **Hard water** reacts with soap to form scum. This means that more soap is needed to form lather with a given volume of hard water than with the same volume of soft water. Soapless detergents never form scum.

You can use soap solution to compare the hardness of different water samples. The more soap solution needed to form a permanent lather (one that doesn't disappear on shaking), the harder the water.

## What makes water hard?

Hard water contains dissolved compounds, usually of calcium or magnesium. The compounds are dissolved when water flows through limestone or chalk rock.

Soft water does not contain dissolved calcium or magnesium compounds.

There are two types of hard water:
- **Permanent hard water** remains hard, even when it is boiled. It contains dissolved calcium and sulfate ions.
- **Temporary hard water** is softened by boiling.

---

**H**   Temporary hard water contains **hydrogen carbonate ions ($HCO_3^-$)**. On heating, these ions decompose to produce carbonate ions ($CO_3^{2-}$). The carbonate ions react with calcium ions ($Ca^{2+}$) or magnesium ions ($Mg^{2+}$) in the water to make calcium carbonate or magnesium carbonate.

$$\text{calcium hydrogen carbonate} \rightarrow \text{calcium carbonate} + \text{carbon dioxide} + \text{water}$$

$$Ca(HCO_3)_2 \, (aq) \rightarrow CaCO_3 \, (s) + CO_2 \, (g) + H_2O \, (l)$$

Calcium carbonate and magnesium carbonate are insoluble in water, so they form as precipitates. The precipitates form scale in kettles and boilers. The boiled water contains few calcium or magnesium ions. It has been softened.

---

## Exam tip    AQA

Don't forget the difference between temporary and permanent hardness – temporary hard water is softened by boiling, permanent hard water remains hard when it is boiled.

### Revision objectives

- ✔ explain what makes water hard or soft
- ✔ describe how to measure water hardness
- ✔ evaluate the environmental, social and economic aspects of water hardness
- ✔ describe and explain three water-softening methods

### Student book references

**3.7**   Water – hard or soft?

**3.8**   Making hard water soft

**3.9**   Water for life

### Specification key

✔ C3.2.1

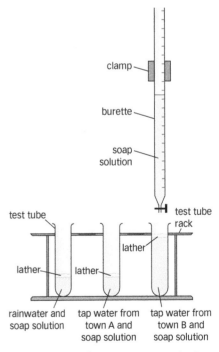

▲ The water from town B needs the most soap solution to form lather. This shows that the water from town B is harder than water from town A, and harder than rainwater.

## Key words

soft water, hard water, hydrogen carbonate ion ($HCO_3^-$), sodium carbonate, ion exchange column

# Hard and soft water – costs and benefits

Hard water is good for health. Its calcium compounds help with the development and maintenance of teeth and bones.

Calcium compounds also help to reduce heart disease.

Hard water also has disadvantages:
- More soap is needed for washing, since scum is formed. This increases costs.
- Heating temporary hard water in kettles and boilers produces scale. Scale makes kettles and heating systems less efficient. This increases costs and may increase greenhouse gas emissions.

## Softening hard water: sodium carbonate

You can soften hard water by adding **sodium carbonate**. Sodium carbonate is soluble in water. In hard water, its carbonate ions react with dissolved calcium and magnesium ions. Calcium carbonate and magnesium carbonate form as precipitates. They can be removed by filtering.

The equations below summarise one of these reactions:

calcium ions + carbonate ions → calcium carbonate

$$Ca^{2+} (aq) \quad + \quad CO_3^{2-} (aq) \quad \rightarrow \quad CaCO_3 (s)$$

## Softening hard water: ion exchange

**Ion exchange columns** swap calcium and magnesium ions with sodium or hydrogen ions.

In the ion exchange column on the left, sodium ions are attached to the resin. Water flows through the column from the top. As the water flows down, calcium and magnesium ions from the water stick to the resin. Sodium ions from the resin dissolve in the water.

After a while the ion exchange column stops working – all its sodium ions have been replaced by calcium and magnesium ions. You then need to pour sodium chloride solution through the column. Sodium ions from the solution stick to the resin. Calcium and magnesium ions are flushed away. The column is ready to use again.

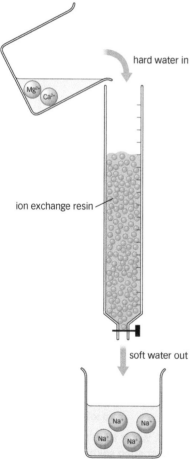

hard water in

ion exchange resin

soft water out

▲ This ion exchange column softens hard water. Ions are not drawn to scale.

---

### Questions

1   Explain what makes water hard.

2   Identify the advantages and disadvantages of hard and soft water.

3   List **three** ways of softening hard water.

# Questions
## Water

### Working to Grade E

1   In the list below, write **S** next to the statements that are true for soft water. Write **H** next to the statements that are true for hard water.
    a   It contains compounds that help bones develop.
    b   It needs only a little soap to make lather.
    c   When heated it may produce scale in kettles.
    d   It helps reduce heart disease.
    e   It contains compounds that help maintain healthy teeth.

2   The table shows three steps taken by water companies to produce drinking water. Draw lines to match each step to one reason.

| Step | Reason |
|---|---|
| Choose an appropriate source | to remove solids from the water. |
| Pass the water through filter beds | to kill bacteria in the water. |
| Add chlorine | to minimise the treatment needed to make the water safe to drink. |

3   Use the words in the box below to complete the sentences that follow. Each word may be used once, more than once, or not at all.

    scale   detergents   less   lather   more   scum

    Soft water easily forms _____ with soap. Hard water reacts with soap to form _____ so _____ soap is needed to form lather. Soapless _____ do not form scum.

4   Put ticks next to the compounds that may be dissolved in hard water to make the water hard.
    a   calcium sulfate
    b   sodium sulfate
    c   calcium hydrogen carbonate
    d   magnesium carbonate

### Working to Grade C

5   In some areas, water companies add fluoride compounds to water. Identify **two** arguments for adding fluorides to water, and **two** arguments against adding fluorides to water.

6   Describe **one** difference between permanent hard water and temporary hard water.

7   Martha investigates the hardness of water from three villages. She takes three 10 cm³ samples of water from each village. She measures the number of drops of soap solution required to form permanent lather. Her results are in the table below.

| Village | Number of drops of soap solution needed to make permanent lather | | | |
|---|---|---|---|---|
| | Run 1 | Run 2 | Run 3 | Mean |
| A | 2 | 3 | 4 | |
| B | 23 | 17 | 20 | |
| C | 43 | 45 | 29 | |

a   Draw a ring around the anomalous result in the table.
b   Calculate the missing means and write them in the table. Ignore the anomalous result.
c   Suggest why Martha tested three samples of water from each village.
d   Write the letters of the villages in order of increasing water hardness.
e   In which village would you expect little scale to be formed in kettles?
f   Which village or villages are most likely to obtain their water from an underground source in a limestone area?

8   This diagram shows how an ion exchange column makes hard water soft. Write the letters of the labels below in the correct boxes. You may write one or more letters in each box.

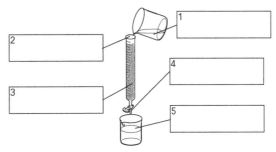

A   Hard water enters the ion exchange column here.
B   After use, calcium or magnesium ions are attached to this.
C   Soft water leaves the ion exchange column here.
D   This water contains dissolved calcium or magnesium ions.
E   This water contains dissolved sodium ions.
F   Before use, sodium ions are attached to this.

9   Explain how adding sodium carbonate softens hard water.

10  Pure water can be produced from seawater by distillation. Explain why, in the UK, the process is very expensive.

### Working to Grade A*

11  The statements below describe why temporary hard water forms scale in kettles.
    The statements are in the wrong order. Write the letters of the statements in the best order in the boxes at the end of this question.
    A   These ions decompose on heating.
    B   These deposit in kettles as scale.
    C   They react with calcium and magnesium ions.
    D   Temporary hard water contains hydrogen carbonate ions.
    E   Carbonate ions are produced.
    F   Precipitates are formed.

|  |  |  |  |  |  |
|---|---|---|---|---|---|
|  |  |  |  |  |  |

1 Some friends are discussing the advantages and disadvantages of adding chlorine to drinking water.

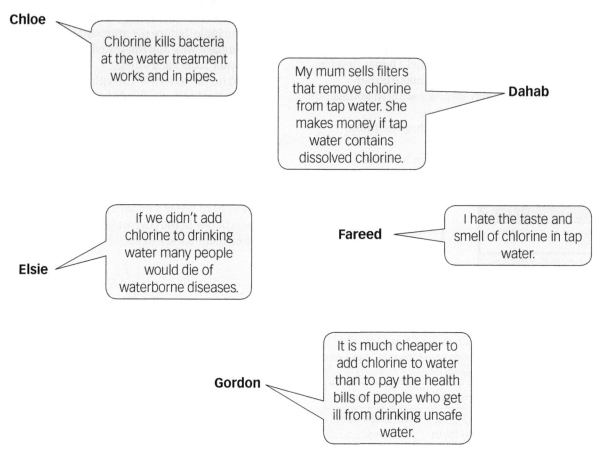

**Chloe**: Chlorine kills bacteria at the water treatment works and in pipes.

**Dahab**: My mum sells filters that remove chlorine from tap water. She makes money if tap water contains dissolved chlorine.

**Elsie**: If we didn't add chlorine to drinking water many people would die of waterborne diseases.

**Fareed**: I hate the taste and smell of chlorine in tap water.

**Gordon**: It is much cheaper to add chlorine to water than to pay the health bills of people who get ill from drinking unsafe water.

Use your knowledge and understanding, and the opinions above, to evaluate the issues involved in adding chlorine to drinking water.

*In this question you will get marks for using good English, organising information clearly, and using scientific words correctly.*

.........................................................................................................................................................................

.........................................................................................................................................................................

.........................................................................................................................................................................

.........................................................................................................................................................................

.........................................................................................................................................................................

.........................................................................................................................................................................

.........................................................................................................................................................................

.........................................................................................................................................................................

.........................................................................................................................................................................

.........................................................................................................................................................................

*(6 marks)*
*(Total marks: 6)*

**2** A student wants to compare the hardness of water in his home on different days.

**a** Why might the hardness of the water be different on different days?

Tick the **two** best answers.

On different days, the water company adds different amounts of chlorine to the water. ☐

The student has a water softener at home, but it doesn't work when the column is saturated with calcium ions. ☐

On different days, the water company supplies water from different sources. ☐

On warmer days the water dissolves more metal ions as it flows through the pipes. ☐

*(2 marks)*

**b** The student uses the apparatus below to test water samples collected on different days.

Describe how the student could use the apparatus to compare the hardness of water samples collected on different days.

........................................................................

........................................................................

........................................................................

........................................................................

........................................................................

*(3 marks)*

clamp

soap solution

burette

test tube

test tube rack

**c** The student's results are in the table below.

| Date | Number of drops of soap solution needed to make permanent lather |
|---|---|
| 20 March | 10 |
| 3 April | 2 |
| 23 August | 3 |
| 2 October | 59 |

**i** From the results in the table, identify the date on which the hardest water sample was collected.

........................................................................

*(1 mark)*

**ii** Suggest how the student could improve his investigation.

........................................................................

*(1 mark)*

**iii** On which of the dates in the table would the student expect the most scale to be formed in his kettle?

........................................................................

*(1 mark)*

**(Total marks: 8)**

3  The diagram shows an ion exchange column used to soften hard water. Before use, sodium ions are attached to the ion exchange resin.

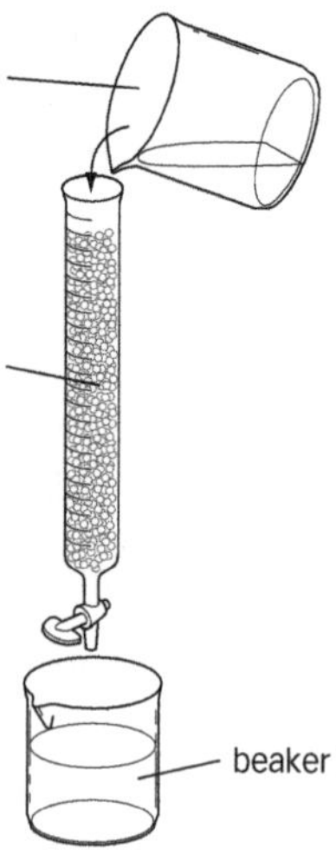

**a** **i**  Name the **two** ions that may make water hard.

.......................................... and ..........................................

*(2 marks)*

**ii**  Compare the water in beaker B with that in beaker A.

..............................................................................................................

..............................................................................................................

..............................................................................................................

*(2 marks)*

beaker A

ion exchange resin

beaker B

**b**  Evaluate the advantages and disadvantages of using a home water softener.

.........................................................................................................................................

.........................................................................................................................................

.........................................................................................................................................

*(3 marks)*

**(Total marks: 7)**

4  This question is about temporary and permanent hard water.

**a**  Describe **one** difference between temporary hard water and permanent hard water.

.........................................................................................................................................

*(1 mark)*

**b**  Adding sodium carbonate softens both temporary and permanent hard water. Explain how.

.........................................................................................................................................

.........................................................................................................................................

*(2 marks)*

**H**

**c**  Describe and explain one method of softening temporary hard water that does not involve adding a chemical to the water, and does not involve using an ion exchange resin.

.........................................................................................................................................

.........................................................................................................................................

.........................................................................................................................................

.........................................................................................................................................

*(4 marks)*

**(Total marks: 7)**

# 6: Calculating and explaining energy change   C3

## Measuring energy changes when fuels burn

Burning reactions are **exothermic**. So when fuels burn, energy is released (given out). Different fuels release different amounts of energy. Energy is measured in **joules** (J). One thousand joules is one **kilojoule**, kJ.

You can use the apparatus below to compare the energy released when different fuels burn. The technique is called **calorimetry**.

To find the energy transferred to the water by a burning fuel, follow the steps below:
- Find the total mass of the spirit burner + fuel.
- Measure out a known mass of water.
- Record the water temperature.
- Burn the fuel so it heats the water.
- Record the new water temperature.
- Find the new mass of the spirit burner + fuel.
- Use the equation below to calculate the amount of energy transferred to the water:

$$\text{energy (in J)} = \text{mass of water (in g)} \times \text{specific heat capacity of water (in J/g°C)} \times \text{temperature change (in °C)}$$

$$Q = mc\Delta T$$

- Divide the value of $Q$ by the mass of fuel burnt. This is the energy released by 1 g of fuel.

Not all the energy released on burning is transferred to the water. Some is transferred to the calorimeter, some to the rest of the apparatus, and some to the air.

### Revision objectives
- work out the energy released on burning different fuels
- calculate the energy released by reactions in solution

### Student book references
**3.12** Measuring food and fuel energy
**3.13** Energy changes

### Specification key
✔ C3.3.1 a – c

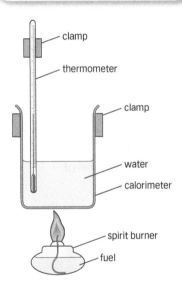

clamp, thermometer, clamp, water, calorimeter, spirit burner, fuel

### Worked example

George uses two fuels – methanol and ethanol – to heat 100 g of water. He collects the data in the table.

The specific heat capacity of water = 4.2 J/g°C

How much energy does each fuel transfer to the water?

| | Methanol | Ethanol |
|---|---|---|
| mass of spirit burner + fuel before heating (g) | 260.0 | 270.0 |
| mass of spirit burner + fuel after heating (g) | 259.0 | 269.5 |
| mass of fuel burnt (g) | 1.0 | 0.5 |
| mass of water (g) | 100 | 100 |
| temperature of water before heating (°C) | 20 | 21 |
| temperature of water after heating (°C) | 72 | 50 |
| increase in water temperature (°C) | 52 | 29 |

### Exam tip   AQA

When calculating the energy released on burning fuels, remember that the '*m*' in the equation $Q = mc\Delta T$ is the mass of the water, not the mass of fuel that was burnt.

For 1 g of methanol, energy transferred = $m \times c \times \Delta T$
$$= 100 \times 4.2 \times 52$$
$$= 21\,840\,J$$
For 0.5 g of ethanol energy transferred = $m \times c \times \Delta T$
$$= 100 \times 4.2 \times 29$$
$$= 12\,180\,J$$
So for 1 g of ethanol, energy transferred = $12\,180 \div 0.5$
$$= 24\,360\,J$$
So ethanol transferred more energy to the water per gram of fuel.

clamp

test tube

water, 10 g

hold the handle here

needle

burning crisp

wooden handle

heat resistant mat

## Measuring energy changes when foods burn

You can use the apparatus on the left to measure the energy released when foods burn.

## Measuring energy changes for reactions in solution

You can calculate the energy released or absorbed by a chemical reaction in solution by measuring the temperature change.

## Worked example

Zoë wants to measure the energy transfer for the reaction of a metal powder with a dilute acid. She sets up the apparatus below. She then:

- measures the temperature of the acid
- adds the metal powder, with stirring
- records the maximum temperature reached.

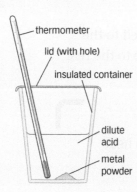

thermometer

lid (with hole)

insulated container

dilute acid

metal powder

Zoë collects the data in the table.

| | |
|---|---|
| volume of acid (g) | 100 |
| temperature at start (°C) | 19 |
| highest temperature reached (°C) | 71 |
| temperature change (°C) | 52 |

She uses the equation $Q = mc\,\Delta T$ to calculate the energy transfer. She assumes that it is only the water in the solution that is being heated.
$$Q = m \times c \times \Delta T$$
$$Q = 100 \times 4.2 \times 52$$
$$Q = 2184\,J$$
So the energy change for the reaction is −2184 J. The negative sign shows that the reaction is exothermic.

The same method can be used to measure the energy transferred in any reactions of solids with water or solutions, and in neutralisation reactions.

## Questions

1   Give the name and symbol of the unit for energy.

2   Calculate the energy transferred when 1 g of a fuel increases the temperature of 100 g of water by 70 °C.
    The specific heat capacity of water, $c$, has a value of 4.2 J/g °C.

3   When measuring the energy transferred from a burning fuel to a container of water, why is it important for the container to be insulated?

## Energy-level diagrams

**Energy-level diagrams** show the relative energies of reactants and products, and the overall energy change of a reaction.

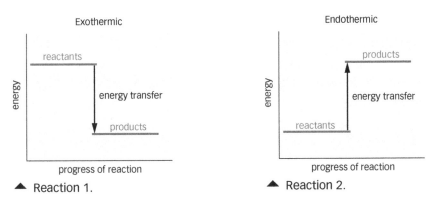

▲ Reaction 1.    ▲ Reaction 2.

The energy-level diagram for reaction 1 shows that the energy stored in the products is less than the energy stored in the reactants. The reaction is exothermic.

The diagram for reaction 2 shows that the reaction is endothermic. The products store more energy than the reactants. Energy changes for endothermic reactions have positive values.

## Activation energy

Reactions can only happen when reactant particles collide. Only particles with enough energy can actually react when they do collide. This means that chemical reactions need energy to get them started. The minimum energy needed to start a reaction is the **activation energy**. Every reaction has its own activation energy.

## Catalysts

Catalysts provide a different pathway for a chemical reaction. The new pathway has a lower activation energy.

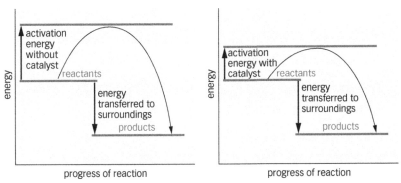

▲ The two energy-level diagrams are for the same reaction. The diagram on the right shows that the activation energy is lower when a catalyst is used.

### Revision objectives

✔ use and interpret energy-level diagrams

✔ explain energy changes in terms of making and breaking bonds

### Student book references

**3.14** Energy-level diagrams

**3.15** Bond breaking, bond making

### Specification key

✔ C3.3.1 d – h

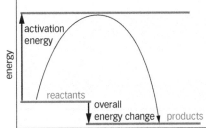

▲ This energy-level diagram shows the overall energy change, and the activation energy, for a reaction. The curved arrow shows the energy as the reaction proceeds.

### Exam tip

In energy-level diagrams:
- if the reactants are higher than the products, the reaction is exothermic
- if the reactants are lower than the products, the reaction is endothermic

## Key words

activation energy, bond energy

**H**

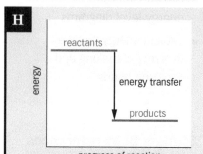

▲ The energy released on forming new bonds is greater than the energy needed to break existing bonds. The reaction is exothermic.

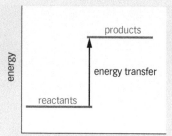

▲ The energy needed to break existing bonds is greater than the energy released from forming new bonds. The reaction is endothermic.

## Questions

1   Sketch an energy-level diagram for an exothermic reaction. Label the axes, and the reactants and products.

2   Explain how a catalyst increases the rate of a reaction.

3   **H** Use ideas about bond energies to explain why the combustion reaction of hydrogen is exothermic. The equation for the reaction is

$$2H_2 + O_2 \rightarrow 2H_2O$$

# Bond breaking and bond making

In a chemical reaction:
*   energy must be supplied to break bonds – endothermic
*   energy is released when new bonds are formed – exothermic.

For example: hydrogen + chlorine→hydrogen chloride

$$H_2 \quad + \quad Cl_2 \quad \rightarrow \quad 2HCl$$

In this reaction, energy must be supplied to break H–H bonds in hydrogen molecules and to break Cl–Cl bonds in chlorine molecules. Energy is released when a hydrogen atom joins with a chlorine atom to make the bond in a hydrogen chloride molecule, H–Cl.

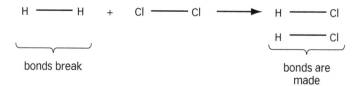

**H** # Endothermic or exothermic?

The difference between the energy needed to break bonds in reactants and the energy released on making new bonds in products determines whether a reaction is exothermic or endothermic.

Every bond has its own **bond energy**, the energy needed to break it. You can use bond energies to calculate energy changes in reactions.

| Bond | Bond energy (kJ/mol) |
|------|----------------------|
| H–H  | 436 |
| F–F  | 158 |
| H–F  | 562 |

**Worked example**

Calculate the energy change for the reaction

$$H_2\ (g) + F_2\ (g) \rightarrow 2HF\ (g)$$

The energy needed to break one mole of H–H bonds and one mole of F–F bonds is (436 + 158) = 594 KJ

Two moles of HF form in the reaction. The energy released when its bonds form is (2 × 562) = 1124 kJ

Overall energy transfer
= energy supplied to break bonds – energy released on making bonds

= 594 – 1124 = –530 KJ

## Hydrocarbon fuels – pros and cons

Most cars are fuelled by petrol or diesel. These fuels are hydrocarbons. They are obtained from crude oil. They burn in car engines to supply the energy needed to make cars move.

Using hydrocarbon fuels has many consequences.

| Category | Consequences |
|---|---|
| Social | • Travel by car can be convenient.<br>• Increasing car use may make towns less pleasant for pedestrians and cyclists.<br>• Supplies of crude oil will one day run out – we cannot rely on petrol and diesel for ever. |
| Economic | • Oil companies make money by selling fuels.<br>• When fuel prices increase people find them less affordable. |
| Environmental | • Burning hydrocarbons makes carbon dioxide, which causes climate change.<br>• Burning diesel makes particulates, which may lead to asthma, lung cancer, and heart disease.<br>• Burning hydrocarbons at high temperature makes oxides of nitrogen. These destroy ozone in the upper atmosphere. Ozone protects us from cancer-causing ultraviolet radiation. |

## Hydrogen fuel

Engineers are developing hydrogen-fuelled cars. The hydrogen can be:

- burnt in combustion engines

$$\text{hydrogen} + \text{oxygen} \rightarrow \text{water}$$
$$2H_2\,(g) \;+\; O_2\,(g) \;\rightarrow 2H_2O\,(l)$$

- used to generate electricity in fuel cells.

When hydrogen-powered vehicles are moving, they do not produce carbon dioxide. However, the hydrogen must be manufactured. It is often made by reacting methane with water. The process produces carbon dioxide, a greenhouse gas.

The table shows some advantages and disadvantages of hydrogen fuel cells compared to burning hydrogen in an internal combustion engine.

| Fuel cells | Hydrogen-fuelled internal combustion engines (ICE) |
|---|---|
| more efficient than ICEs | less efficient than fuel cells |
| batteries expensive | technology well understood |
| include expensive platinum catalysts | nitrogen and oxygen react in the engine to produce nitrogen oxides as well as water |
| few fuel stations supply hydrogen gas for refuelling | |

### Exam tip

If you are asked to evaluate the consequences of burning different fuels, try grouping the consequences into categories, for example, environmental, social, and economic.

### Questions

1  Describe **two** ways in which hydrogen is used as a fuel.

2  Identify **three** consequences of burning hydrocarbon fuels in cars.

## Working to Grade E

1   Write **T** next to the statements that are true. Write corrected versions of the statements that are false.
   a   Energy is normally measured in joules.
   b   100 J = 1 KJ
   c   In the equation $Q = mc\Delta T$, $\Delta T$ represents energy change.
   d   In an energy-level diagram, a curved arrow shows the energy as the reaction proceeds.
   e   In a chemical reaction, energy is released when bonds break.
   f   In a chemical reaction, energy must be supplied to form new bonds.
   g   A catalyst provides a different pathway for a reaction. The new pathway has a higher activation energy.

2   Look at the energy-level diagram.

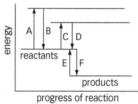

   a   Give the letter of the arrow that shows the overall energy change of the reaction.
   b   Give the letter of the arrow that shows the activation energy for the reaction without a catalyst.
   c   Give the letter of the arrow that shows the activation energy for the reaction when a catalyst is used.
   d   Is the reaction exothermic or endothermic? Explain how you decided.

## Working to Grade C

3   Use data from the table to help you answer the questions below.

| Fuel | State at room temperature | Energy released on burning fuel (kJ/g) |
|---|---|---|
| methane | gas | 56 |
| ethanol | liquid | 30 |
| hydrogen | gas | 143 |

   a   Per gram, which fuel transfers most energy on burning?
   b   Identify some advantages and disadvantages of using hydrogen as a fuel compared to the other fuels in the table.

4   Eduardo uses a burning crisp to heat 100 g of water. The temperature of the water rose by 35 °C. Use the equation $Q = mc\Delta T$ to calculate the energy released by the burning crisp. The specific heat capacity of water is 4.2 J/g°C.

5   Clarissa adds 50 cm³ of sodium hydroxide solution to 50 cm³ of hydrochloric acid solution in an insulated container. The temperature of the water increases from 20 °C to 27 °C.

   a   Calculate the energy change for the reaction.
   b   Is the reaction exothermic or endothermic? Explain how you decided.

6   Use the data on the energy-level diagrams below to answer the questions that follow.

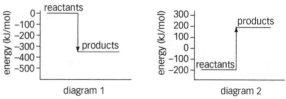

diagram 1                    diagram 2

   a   Give the overall energy change, including a + or − sign for:
      i    the reaction shown in diagram 1
      ii   the reaction shown in diagram 2.
   b   Which of the reactions shown on the energy-level diagrams releases energy? Explain how you decided.

7   For each reaction below, draw every bond that breaks and every bond that is made in the correct boxes in the table.
   a   $H_2 (g) + Cl_2 (g) \rightarrow 2HCl (g)$
   b   $2H_2 (g) + O_2 (g) \rightarrow 2H_2O (g)$
   c   $CH_4 (g) + Cl_2 (g) \rightarrow CH_3Cl (g) + HCl (g)$

| Reaction | Bonds that break | Bonds that are made |
|---|---|---|
| a | | |
| b | | |
| c | | |

## Working to Grade A*

8   The table shows the energy needed to break some covalent bonds. Use data from the table to answer the questions below it.

| Bond | Bond energy (kJ/mol) | Bond | Bond energy (kJ/mol) |
|---|---|---|---|
| H–H | 436 | H–Cl | 432 |
| O=O | 497 | C–H | 413 |
| H–O | 463 | C–Cl | 339 |
| Cl–Cl | 243 | | |

   a   Which bond needs most energy to break it?
   b   Which bond gives out least energy when it is made?
   c   Which bond is strongest?

9   Calculate energy changes for the reactions below. To help you, use your answers to question 7 and the data in question 8.
   a   $H_2 (g) + Cl_2 (g) \rightarrow 2HCl (g)$
   b   $2H_2 (g) + O_2 (g) \rightarrow 2H_2O (g)$
   c   $CH_4 (g) + Cl_2 (g) \rightarrow CH_3Cl (g) + HCl (g)$

**1** A student does an experiment to compare the energy stored in two types of breakfast cereal, WheetyWheels and RicyRings.

He sets up the apparatus below.

He burns 1 g of WheetyWheels, and uses it to heat 100 g of water.

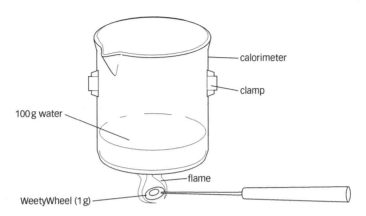

For WheetyWheels, the student obtains the data in the table.

|  | **Run 1** | **Run 2** | **Run 3** |
|---|---|---|---|
| Initial water temperature (°C) | 19 | 20 | 22 |
| Final water temperature (°C) | 51 | 56 | 60 |
| Temperature change (°C) | 32 | 36 | 38 |

**a** Suggest how the student could ensure his investigation was fair.

.................................................................................................................................................................

*(2 marks)*

**b** Use the data in the table to calculate the mean temperature change for WheetyWheels.

Show clearly how you work out your answer.

.................................................................................................................................................................

Mean temperature change = ............°C

*(1 mark)*

**c** Calculate how much energy is transferred to the water by 1 g of WheetyWheels. Use the equation

$$\text{energy released} = \text{mass of water} \times 4.2 \times \text{mean temperature change}$$
$$\text{(in J)} \qquad\qquad \text{(in g)} \qquad\qquad\qquad \text{(in °C)}$$

Show clearly how you work out your answer.

.................................................................................................................................................................

.................................................................................................................................................................

Energy released on burning 1 g of WheetyWheels = ............ J

*(2 marks)*

**d** The student then uses 1 g of burning RicyRings to heat the water.

He calculates that 10 000 J is transferred to the water.

He compares his value to the value on the RicyRings packet.

The packet states that burning 1 g of RicyRings releases 16 000 J.

**i** Suggest **one** reason for the difference between the student's value and the value on the RicyRings packet.

..................................................................................................................................................................

*(1 mark)*

**ii** Suggest **one** change the student could make to his apparatus to improve the data he collects.

..................................................................................................................................................................

*(1 mark)*

*(Total marks: 7)*

**2** A farmer plans to collect the methane gas produced by decaying cow manure.

The methane gas will be piped to local homes and burnt to provide energy for cooking and heating.

Methane is a greenhouse gas.

The equation below shows the products formed when methane burns.

$$\text{methane} + \text{oxygen} \rightarrow \text{carbon dioxide} + \text{water}$$
$$CH_4 + 2O_2 \rightarrow CO_2 + 2H_2O$$

The equation can also be written showing the structural formulae.

**a** **i** Identify two environmental impacts of the farmer's scheme.

**1** ..................................................................................................................................................................

**2** ..................................................................................................................................................................

*(2 marks)*

**ii** Suggest an economic benefit of the scheme.

..................................................................................................................................................................

*(1 mark)*

**b** The combustion (burning) of methane is an exothermic reaction.

Explain how the energy-level diagram below shows that the reaction is exothermic.

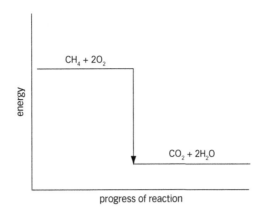

.......................................................................................................................................................................

.......................................................................................................................................................................

.......................................................................................................................................................................

.......................................................................................................................................................................

*(2 marks)*

**H**

**c** Use the bond energies in the table below, and the equations at the start of question 2, to calculate the energy change for the combustion of methane.

Show clearly how you work out your answer.

| Bond | Bond energy (kJ/mol) |
|------|----------------------|
| C=O | 743 |
| O=O | 497 |
| H–O | 463 |
| C–H | 413 |

.......................................................................................................................................................................

.......................................................................................................................................................................

.......................................................................................................................................................................

.......................................................................................................................................................................

Energy released on burning one mole of methane = ............ kJ

*(4 marks)*

*(Total marks: 9)*

## Presenting data and using data to draw conclusions

In this module there are several opportunities to present data and draw conclusions. These include presenting data and drawing conclusions on water hardness, softening water, and energy changes in chemical reactions.

As well as demonstrating these skills practically, you are likely to be asked to comment on data presented by others, and on the conclusions others draw from data. The examples below offer guidance in these skill areas. They also give you the chance to practise using your skills to answer the sorts of questions that may well come up in exams.

## Comparing the hardness of water from different towns

1   Irma did tests to compare the hardness of water from five towns. She took $10\,cm^3$ water from each town. She added soap solution to each sample, and counted the number of drops required to make a permanent lather. For each town, she repeated the test three times.

Her results are in the table.

| Town | Number of drops of soap solution required for permanent lather | | | |
|---|---|---|---|---|
| | Run 1 | Run 2 | Run 3 | Mean |
| A | 2 | 5 | 8 | |
| B | 28 | 26 | 30 | |
| C | 6 | 6 | 9 | |
| D | 58 | 62 | 54 | |
| E | 6 | 7 | 8 | |

a   Of all the towns in the table, identify the town that has the greatest range for the number of drops of soap solution required.

The range of a data set refers to the maximum and minimum values.
- For town A the range is 2 to 8 (a difference of 6).
- For town B the range is 26 to 30 (a difference of 4).

So of the two towns above, town A has the greater range.

b   Calculate the mean number of drops of soap solution required for each town.

The mean of the data refers to the sum of all the measurements divided by the number of measurements taken.
- For town A the mean = $(2 + 5 + 8) \div 3 = 5$
- For town B the mean = $(28 + 26 + 30) \div 3 = 28$

c   Would it be better to draw a line graph or a bar chart to display the data in the table? Give a reason for your decision.

- If one of the variables is categoric (it can be described by a label, and has no numerical meaning), use a bar chart to display the data.
- If both the dependent and independent variables are continuous (can have any numerical value), use a line graph to display the data.

d   Draw a bar chart or a line graph to display the data in the table.

When drawing a bar chart or line graph, remember the following:
- Label both axes clearly. Don't forget to include units, if they are needed.
- Choose sensible scales.
- Plot the data accurately.

## Softening water

Now follow the examiner's advice given for the questions above by answering the exam question below.

2   Jake investigated the effect of adding different amounts of sodium carbonate to $10\,cm^3$ samples of hard water from the same source.

A summary of his results is in the table.

| Volume of sodium carbonate solution added ($cm^3$) | Number of drops of soap solution required for permanent lather |
|---|---|
| 1 | 38 |
| 2 | 29 |
| 3 | 21 |
| 4 | 10 |
| 5 | 2 |
| 6 | 2 |
| 7 | 15 |
| 8 | 2 |

a   Draw a line graph to display the data in the table.
b   Use your graph to identify any anomalous values in the data set.

An anomalous value is one that does not fit the pattern shown by the rest of the data. Displaying data on a line graph may help you to spot anomalous data.

c   Write a sentence or two to describe the relationship shown on your graph.

# AQA Upgrade

## Answering an extended writing question

QUESTION

*In this question you will be assessed on using good English, organising information clearly, and using specialist terms where appropriate.*

**1** The two energy-level diagrams below are for the same reaction. Explain and compare what the two diagrams show, suggesting reasons for any differences. In your answer, refer to the labelled arrows. *(6 marks)*

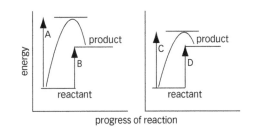

---

they both have the same overall energie change and the arrows of some are the same and some are different and there is a curved lighn and some strayt lighns but i dont no why.

**G–E**

**Examiner:** This answer is typical of a grade-G candidate. It is worth just one mark, gained for recognising that the overall energy change is the same in both diagrams.

The candidate has used one specialist term. There is no punctuation, and there are several spelling mistakes.

---

The reactions is endothermic. I knows this becuase the products have more energy than the reactants (arrow B). Arrows C and A are different lenghs this is to do with activation energy – in diagram C the activation energy is less.
Energy is measured in joules and hydrogen can be burnt as a fuel in combustion engines.

**D–C**

**Examiner:** This answer is worth three marks out of six. It is typical of a grade-C or -D candidate.

The candidate has correctly pointed out that the reaction is endothermic, and referred to three of the four arrows. The candidate has recognised that arrows C and A refer to activation energy, but has not suggested a reason for the different activation energy values.

The answer is well organised, with a few mistakes of grammar and punctuation. There are two spelling mistakes. The last part of the answer is not relevant.

---

Both diagrams show that the relative energy of the products is greater than that of the reactants. So the reaction is endothermic.
Arrows B and D show the same difference in energy. They represent the overall energy change of the reaction. Since both diagrams represent the same reaction, the overall energy change shown on both diagrams is the same.
Arrows A and C represent activation energy. The activation energy on diagram 1 (arrow A) is greater than that on diagram 2 (arrow C). This could be because a catalyst has been used in the reaction represented by diagram 2. Catalysts provide a different pathway for a reaction that has a lower activation energy.

**B–A\***

**Examiner:** This is a high-quality answer, typical of an A* candidate. It is worth six marks out of six.

The candidate has explained and compared the features of the energy-level diagrams very clearly, and referred to the labelled arrows. The reason given for the difference in the two diagrams is correct, and explained in detail.

The answer is well organised. The spelling, punctuation, and grammar are faultless. The candidate has used several specialist terms.

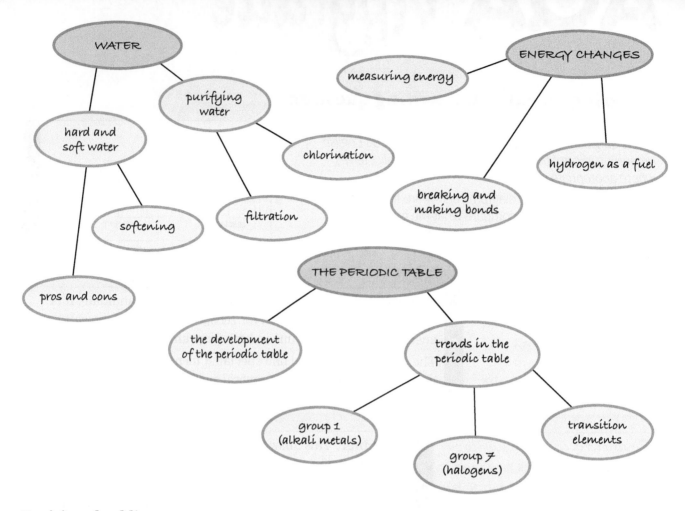

## Revision checklist

- Newlands and Mendeleev classified the elements by arranging them in atomic weight order.
- Mendeleev overcame problems in his periodic table by leaving gaps for undiscovered elements.
- In the modern periodic table, elements are arranged in atomic number order.
- Group 1 elements are low-density metals that react with non-metals to form ionic compounds, and with water to release hydrogen. Their hydroxides form alkaline solutions.
- Compared to the Group 1 elements, the transition elements are stronger, harder, have higher melting points, and are less reactive. They have ions with different charges, form coloured compounds, and are useful as catalysts.
- The Group 7 elements react with metals to form ionic compounds. More reactive halogens displace less reactive halogens from aqueous solutions of their salts.
- Hard water contains dissolved compounds of calcium or magnesium. Calcium compounds are good for heart health, teeth, and bones.
- Hard water increases costs because more soap is needed, and scale makes kettles and heating systems less efficient.

- Temporary hard water is softened by boiling. Permanent hard water remains hard when it is boiled.
- Hard water can be softened by adding sodium carbonate or by passing it through ion exchange columns.
- Water is made safe to drink by filtering, and by adding chlorine to reduce microbes. Adding fluoride may improve dental health.
- The energy change of a chemical reaction can be calculated using the equation $Q = mc\Delta T$. $\Delta T$ is the temperature change of water heated by a burning fuel, or of reacting solutions.
- Energy-level diagrams show the relative energies of reactants and products, the activation energy, and the overall energy change of a reaction.
- In chemical reactions, energy is supplied to break bonds, and energy is released when new bonds are formed.
- Catalysts reduce the minimum amount of energy needed to start a chemical reaction (the activation energy).
- Hydrogen can be burnt as a fuel in combustion engines, or used in fuel cells to produce electricity to power vehicles.

## Flame tests

You can use a flame test to identify the metal ion in a salt.

- Dip the end of a clean nichrome wire in the salt.
- Hold the end of the wire in a hot Bunsen flame.
- Observe the flame colour.

| Metal ion | Flame colour |
|---|---|
| lithium | crimson |
| sodium | yellow |
| potassium | lilac |
| calcium | red |
| barium | green |

## Using sodium hydroxide to identify other metal ions

Sodium hydroxide solution helps to identify many other metal ions.

- Dissolve the salt in pure water.
- Add a few drops of sodium hydroxide solution.
- Observe the colour of any precipitate formed.

| Metal ion | Metal ion formula | Colour of hydroxide precipitate |
|---|---|---|
| aluminium | $Al^{3+}$ | white – dissolves in excess sodium hydroxide solution to form a colourless solution |
| calcium | $Ca^{2+}$ | white |
| magnesium | $Mg^{2+}$ | white |
| copper(II) | $Cu^{2+}$ | blue |
| iron(II) | $Fe^{2+}$ | green |
| iron(III) | $Fe^{3+}$ | brown |

## Testing for carbonates

- Add a few drops of dilute hydrochloric acid to the solid.
- If it fizzes, a gas is being made.
- Test the gas with limewater. A white precipitate may form, which makes the limewater cloudy. This shows that the gas is carbon dioxide, and the solid is a carbonate.

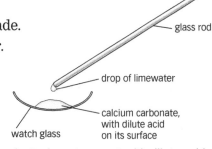

▲ Carbonates react with dilute acids to form carbon dioxide gas.

### Exam tip

You are expected to know all the chemical tests described on this page, and their results. Make sure you learn them!

### Questions

1 Give the colours of these precipitates: copper hydroxide, magnesium hydroxide, and iron(II) hydroxide.

2 Describe the test for aluminium ions, and the results you would expect.

3 Describe the test for carbonates, and the results you would expect.

# and doing titrations

## Revision objectives

- ✔ describe and interpret tests to identify halide and sulfate ions
- ✔ describe how to do a titration
- ✔ use titrations to find the volumes of acids and alkalis that react
- ✔ calculate the concentration of a solution

## Student book references

**3.18** Identifying negative ions

**3.20** Titrations – 1

## Specification key

✔ C3.4.1 e – h

## Key words

end point, rough titration, concentration, solute

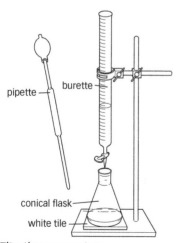

▲ Titration apparatus.

pipette, burette, conical flask, white tile

# Testing for halide ions

Use silver nitrate solution to test for chloride, bromide, and iodide ions.

- Dissolve a little of the solid in dilute nitric acid.
- Add silver nitrate solution.

| Halide ion | Colour of precipitate | Formula of precipitate |
|---|---|---|
| chloride | white | AgCl |
| bromide | cream | AgBr |
| iodide | yellow | AgI |

The equation for the reaction of sodium bromide with silver nitrate is:

$$\text{sodium bromide} + \text{silver nitrate} \rightarrow \text{silver bromide} + \text{sodium nitrate}$$

The ionic equation for the reaction is:

$$Br^-(aq) + Ag^+(aq) \rightarrow AgBr(s)$$

# Testing for sulfate ions

To find out if a solid or solution contains sulfate ions, $SO_4^{2-}$, follow the steps below.

- Dissolve the solid in dilute hydrochloric acid.
- Add barium chloride solution.

If a white precipitate forms, the sample includes a sulfate ion.

The ionic equation for the reaction is:

$$Ba^{2+}(aq) + SO_4^{2-}(aq) \rightarrow BaSO_4(s)$$

# Finding the volumes of acids and alkalis that react

**Titrations** show the volumes of acids and alkali solutions that react.

To find the volume of hydrochloric acid that exactly reacts with $25.00\,cm^3$ of sodium hydroxide solution, follow the steps below.

1. Use a pipette to measure out $25.00\,cm^3$ sodium hydroxide solution.
2. Transfer the solution to a conical flask.
3. Add a few drops of indicator solution to the sodium hydroxide solution in the conical flask. Phenolphthalein indicator will turn pink.
4. Pour hydrochloric acid into a burette. Read the scale.
5. Place the conical flask under the burette. Allow the hydrochloric acid to run into the conical flask. Swirl the mixture in the flask.

6  When the phenolphthalein indicator just turns colourless, the titration has reached its **end point**.
7  Read the scale on the burette. Calculate the volume of acid added. You have now finished the **rough titration**.
8  Repeat steps 1–7. This time, add the acid one drop at a time as you approach the end point. Swirl after every drop. Repeat until you have three consistent values for the acid volume.

## Calculating titration volumes

The table gives titration results for neutralising $25.00\,cm^3$ of sodium hydroxide with $2.00\,mol/dm^3$ hydrochloric acid from a burette.

|  | Rough | Run 1 | Run 2 | Run 3 |
|---|---|---|---|---|
| initial burette reading (cm³) | 0.90 | 13.10 | 25.90 | 0.10 |
| final burette reading (cm³) | 14.00 | 25.90 | 38.60 | 12.90 |
| volume of acid added (cm³) | 13.10 | 12.80 | 12.70 | 12.80 |

You can use the results in the table to calculate the mean volume of acid added. Do not include the value for the rough titration.

Mean volume = $(12.80 + 12.70 + 12.80) \div 3 = 12.77\,cm^3$

### H  Calculating concentration

**Concentration** is the amount of solute per unit volume of solution.

A **solute** is a substance dissolved in solution.

Chemists measure concentration in grams per cubic decimetre $(g/dm^3)$ or in moles per cubic decimetre $(mol/dm^3)$.

One mole (mol) of a substance is its unit mass in grams.

$$\text{concentration } (g/dm^3) = \frac{\text{mass of solute (g)}}{\text{volume of solution } (dm^3)}$$

or

$$\text{concentration } (mol/dm^3) = \frac{\text{number of moles of solute (mol)}}{\text{volume of solution } (dm^3)}$$

H

**Worked example**

Calculate the concentration of a solution that has 0.25 moles of sodium hydroxide in $250\,cm^3$ of solution.

Volume of solution in $dm^3$
$$= \frac{250\,cm^3}{1000\,cm^3}$$
$$= 0.25\,dm^3$$

Concentration in $mol/dm^3$
$$= \frac{\text{number of moles}}{\text{volume in } dm^3}$$
$$= \frac{0.25\,mol}{0.25\,dm^3}$$
$$= 1\,mol/dm^3$$

**Questions**

1  Describe the test for bromide ions, and the result you would expect.

2  Explain why a titration should be carried out several times, until consistent results are obtained.

3  H Calculate the concentration of a solution that has 0.10 mole of acid dissolved in $500\,cm^3$ of solution.

## Revision objectives

- ✔ use titration results to calculate concentrations of solutions

## Student book references

**3.21** Titrations – 2

**3.22** Titrations – 3

## Specification key

✔ C3.4.1 h

## Exam tip  **AQA**

Practise doing as many titration calculations as possible.

## Questions

1  Calculate the number of moles of hydrochloric acid in 25.00 cm³ of 2.00 mol/dm³ solution.

2  Calculate the mass of potassium hydroxide in 24.00 cm³ of 0.2 mol/dm³ solution.

3  Izzy uses 20.7 cm³ of sodium hydroxide solution of concentration 8.0 g/dm³ to neutralise 20.0 cm³ of sulfuric acid. Calculate the concentration of the sulfuric acid.

# Calculating masses and moles

If you know the concentration of a solution, and its volume, you can calculate the mass of solute, or the number of moles of solute, in a given volume of solution.

> ### Worked example
>
> What mass of potassium nitrate is in 500 cm³ of a 202 g/dm³ solution?
>
> Volume of solution in dm³ = $\dfrac{500 \, cm^3}{1000 \, cm^3}$ = 0.5 dm³
>
> Mass = concentration in g/dm³ × volume in dm³
>
> = 202 g/dm³ × 0.5 dm³ = 101 g

# Using titrations to calculate concentrations

If you know the concentration of one reactant, you can use titration results to work out the concentration of the other reactant.

> ### Worked example
>
> Rosie has a solution of sodium hydroxide of unknown concentration. She measures 25.00 cm³ of the sodium hydroxide solution into a conical flask. She titrates the sodium hydroxide solution with sulfuric acid of concentration 0.500 mol/dm³. The volume of sulfuric acid required is 24.5 cm³. Calculate the concentration of the sodium hydroxide solution.
>
> *Calculate the number of moles of sulfuric acid.*
>
> Number of moles = concentration in mol/dm³ × volume in dm³
>
> = 0.500 mol/dm³ × (24.5 ÷ 1000) dm³
>
> = 0.0123 mol
>
> *Write a balanced equation for the reaction. Use it to work out the number of moles of sodium hydroxide in 25.00 cm³ of solution.*
>
> $2NaOH(aq) + H_2SO_4 \, (aq) \rightarrow Na_2SO_4 \, (aq) + 2H_2O \, (l)$
>
> The equation shows that 2 moles of sodium hydroxide react with 1 mole of sulfuric acid.
>
> Rosie has 0.0123 mol of sulfuric acid.
>
> So the number of moles of sodium hydroxide = (0.0123 × 2)
>
> = 0.0246
>
> *Calculate the concentration of sodium hydroxide in mol/dm³*
>
> concentration = $\dfrac{number \; of \; moles}{volume \; in \; dm^3}$ = $\dfrac{0.0246 \, mol}{(25 \div 1000) \, dm^3}$
>
> = 0.984 mol/dm³

# Questions
## Further analysis and quantitative chemistry

### Working to Grade E

1 Complete the table below.

| Compound of... | Flame colour |
|---|---|
| lithium | |
| sodium | |
| potassium | |
| calcium | |
| barium | |

2 Draw lines to match each precipitate to its colour.

| Name of precipitate |
|---|
| aluminium hydroxide |
| copper(ii) hydroxide |
| iron(ii) hydroxide |
| calcium hydroxide |
| iron(iii) hydroxide |
| magnesium hydroxide |

| Colour |
|---|
| blue |
| white |
| brown |
| green |
| white |
| white |

3 Draw a ring around the name of the precipitate that dissolves in excess sodium hydroxide solution.

calcium hydroxide     aluminium hydroxide

magnesium hydroxide

### Working to Grade C

4 Use the words in the box below to complete the sentences that follow. Each word may be used once, more than once, or not at all.

> hydrochloric  white  silver nitrate  sulfuric
> yellow  green  barium chloride  cream
> barium sulfate  nitric

To test for sulfate ions in a solution, add _____ acid and _____ _____ solution. If a _____ precipitate forms, sulfate ions are present in the solution.
To test for chloride ions in a solution, add _____ acid and _____ _____ solution. If a _____ precipitate forms, chloride ions are present.

5 Describe how to find out whether a compound is a carbonate. Include the names of the chemicals you would use and describe the observations you would expect to make.

6 Dan writes out the steps for doing a titration to find the volume of hydrochloric acid that reacts with 25.00 cm³ of sodium hydroxide solution. He makes **one** mistake in **each** step. Write a corrected version of each step.

  a Use a measuring cylinder to measure out exactly 25.00 cm³ of sodium hydroxide solution.

  b Transfer the solution to a beaker.

  c Using a funnel, pour the hydrochloric acid into a burette. Add a few drops of indicator to the burette.

  d Read the scale on the burette. Add hydrochloric acid from the burette to the sodium hydroxide solution in the beaker until the indicator changes colour. Read the scale on the burette and calculate the volume of acid added.

  e Repeat steps **a** to **d** once more.

7 Sabrina does a set of three titrations. Her results are in the table.

| | Rough | Run 1 | Run 2 | Run 3 |
|---|---|---|---|---|
| initial burette reading (cm³) | 0.20 | 10.20 | 9.80 | 19.50 |
| final burette reading (cm³) | 10.20 | 20.00 | 19.50 | 29.30 |
| volume of acid added (cm³) | 10.00 | 9.80 | | 9.80 |

  a Calculate the volume of acid added in run 2.

  b Calculate the mean volume of acid added for runs 1, 2, and 3.

### Working to Grade A*

8 Calculate the concentrations of the solutions in the table. Give your answers in $g/dm^3$.

| Solute | | Volume of solution (cm³) |
|---|---|---|
| Name | Mass of solute in solution (g) | |
| copper sulfate | 8 | 2000 |
| sodium chloride | 15 | 500 |
| magnesium sulfate | 20 | 100 |

9 Calculate the number of moles of sulfuric acid in 25.00 cm³ of a $1.0 \, mol/dm^3$ solution.

10 Calculate the number of moles of sodium hydroxide in 47.00 cm³ of a $0.1 \, mol/dm^3$ solution.

11 Calculate the mass of sodium chloride in 1.00 dm³ of a $1.0 \, mol/dm^3$ solution. The formula of sodium chloride is $NaCl$.

12 Calculate the mass of magnesium chloride in 500.00 cm³ of a $0.50 \, mol/dm^3$ solution. The formula of magnesium chloride is $MgCl_2$.

13 Rob uses 23.10 cm³ of hydrochloric acid of concentration $1.00 \, mol/dm^3$ to neutralise 25.00 cm³ of sodium hydroxide solution. Calculate the concentration of the sodium hydroxide solution. Give your answer in $mol/dm^3$.

14 Mel places 25.00 cm³ of nitric acid in a flask. She uses 13.40 cm³ of $0.10 \, mol/dm^3$ sodium hydroxide solution to neutralise the acid. Calculate the concentration of the nitric acid. Give your answer in $mol/dm^3$.

1 A student has a mixture of two white powders, X and Y.

She does a series of tests to identify the substances in the mixture.

Her results are in the table.

| Test number | Test | Observations |
|---|---|---|
| 1 | Flame test. | Red or crimson flame – not sure which. |
| 2 | Dissolve in pure water. Add sodium hydroxide solution. | White precipitate. Insoluble in excess sodium hydroxide. |
| 3 | Add dilute hydrochloric acid to the mixture of powders. | Fizzed. The bubbles made limewater cloudy. |
| 4 | Dissolve in pure water. Add sulfuric acid and barium chloride solution. | White precipitate. |
| 5 | Dissolve in pure water. Add nitric acid and silver nitrate solution. | White precipitate. |

a   Describe how to do a flame test.

   .................................................................................................................................................

   .................................................................................................................................................
   *(1 mark)*

b   Suggest **three** alternative conclusions the student might make after tests 1 and 2.

   Use evidence from the table to support each possible conclusion.

   .................................................................................................................................................

   .................................................................................................................................................

   .................................................................................................................................................
   *(3 marks)*

c   Which test described in the table is not valid? Explain why.

   .................................................................................................................................................
   *(1 mark)*

d   Name the **two** negative ions that are present in the mixture.

   Give reasons for your decision.

   .................................................................................................................................................

   .................................................................................................................................................
   *(2 marks)*
   *(Total marks: 7)*

**2** Describe how to do a titration to find out the volume of hydrochloric acid that reacts with a given volume of sodium hydroxide solution.

Include the names of each piece of apparatus used.

*In this question you will get marks for using good English, organising information clearly, and using scientific words correctly.*

..........................................................................................................................................................................

..........................................................................................................................................................................

..........................................................................................................................................................................

..........................................................................................................................................................................

..........................................................................................................................................................................

..........................................................................................................................................................................

..........................................................................................................................................................................

..........................................................................................................................................................................

..........................................................................................................................................................................

*(6 marks)*
***(Total marks: 6)***

**3** A student investigated the claim on a carton of blackcurrant drink.

*'Our blackcurrant drink has four times more vitamin C than orange juice.'*

**a** The student first titrated 10.00 cm³ samples of **orange juice** with DCPIP solution.

DCPIP reacts with vitamin C. The more DCPIP required, the greater the amount of vitamin C in the drink sample.

The titration results are in the table.

| | Rough | Run 1 | Run 2 | Run 3 |
|---|---|---|---|---|
| Initial burette reading (cm³) | 1.00 | 13.60 | 25.50 | 37.50 |
| Final burette reading (cm³) | 13.60 | 25.50 | 37.50 | 49.60 |
| Volume of DCPIP added | 12.60 | | | |

**i** Explain why the student did a rough titration.

......................................................................................................................................................................

*(1 mark)*

**ii** Explain why the student repeated the titration several times.

......................................................................................................................................................................

*(1 mark)*

**iii** Use the results in the table to show that the mean volume of DCPIP that reacts with $10\,cm^3$ of orange juice is $12.00\,cm^3$.

You may write in the table above.

......................................................................................................................................................................

*(1 mark)*

**b** The student then titrated $10.00\,cm^3$ samples of the **blackcurrant drink** with DCPIP.

The mean volume of DCPIP required was $0.15\,cm^3$.

Do the results of the titrations with the orange juice and blackcurrant drink support the claim on the blackcurrant drink carton?

Explain your answer.

......................................................................................................................................................................

*(1 mark)*
***(Total marks: 4)***

**H**

**4** Jordan does a titration. He uses $23.00\,cm^3$ of sulfuric acid of concentration $9.8\,g/dm^3$ to neutralise $25.00\,cm^3$ of sodium hydroxide solution.

Calculate the concentration of the sodium hydroxide solution in $mol/dm^3$.

You are advised to write an equation and show all your working.

......................................................................................................................................................................

......................................................................................................................................................................

......................................................................................................................................................................

......................................................................................................................................................................

......................................................................................................................................................................

......................................................................................................................................................................

*(4 marks)*

***(Total marks: 4)***

## Raw materials

**Ammonia** is a compound of nitrogen and hydrogen. Its formula is $NH_3$. Compounds made from ammonia are vital fertilisers.

Ammonia is manufactured by the **Haber process**. The raw materials are its elements:

- Nitrogen is separated from the air.
- Hydrogen may be obtained from natural gas. Methane from the natural gas reacts with water. The products are carbon monoxide and hydrogen:

    methane + water → carbon monoxide + hydrogen

    Hydrogen for the Haber process is sometimes obtained from other sources.

## The Haber process

First, the raw materials for the Haber process are purified. Then the pure nitrogen and hydrogen enter the reaction vessel. In the reaction vessel:

- the gases pass over an iron catalyst
- the temperature is about 450 °C
- the pressure is about 200 atm.

Some of the hydrogen and nitrogen react together to form ammonia.

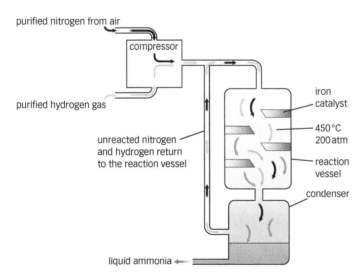

purified nitrogen from air

compressor

purified hydrogen gas

unreacted nitrogen and hydrogen return to the reaction vessel

iron catalyst

450 °C 200 atm

reaction vessel

condenser

liquid ammonia

The Haber process reaction is **reversible**. Whilst some ammonia molecules are being made, others are breaking down to form nitrogen and hydrogen. The ⇌ sign shows that the reaction is reversible.

$$nitrogen + hydrogen \rightleftharpoons ammonia$$
$$N_2 + 3H_2 \rightleftharpoons 2NH_3$$

There are three gases in the reaction vessel – nitrogen, hydrogen, and ammonia.

The mixture of gases moves from the reaction vessel to the condenser. In the condenser, the mixture cools. At –33 °C, ammonia gas condenses to form liquid ammonia. The liquid ammonia is removed.

Nitrogen and hydrogen gases remain in the condenser. They are returned to the reaction vessel.

## Choosing conditions

Ammonia production companies need to maximise their income. They must make as much ammonia as possible, as quickly as possible. They choose conditions for the Haber process that will maximise:

- the **yield** of ammonia (the percentage of ammonia in the equilibrium mixture)
- the rate of the reaction (the speed at which ammonia is produced).

### Pressure

The higher the pressure, the greater the yield of ammonia. But the higher the pressure, the stronger the reaction vessel and pipes need to be, and the more expensive they are. Chemical engineers choose a compromise pressure of about 200 atm – this produces an acceptable yield at a reasonable cost.

### Temperature

The lower the temperature, the higher the yield of ammonia. But at low temperatures the reaction is very slow. Chemical engineers choose a compromise temperature of 450 °C. This produces an acceptable yield at a reasonable rate.

### Energy and environment

Ammonia companies need to keep their energy costs as low as possible. One way of reducing energy requirements is to transfer the energy released as the reaction mixture cools in the condenser to the reaction vessel.

Reducing energy costs helps to reduce the environmental impact of producing ammonia. There are strict laws to prevent companies allowing poisonous ammonia to escape into the air or water.

## Exam tip

Remember the conditions chosen for the Haber process: temperature = 450 °C and pressure = 200 atmospheres (atm).

## Questions

1 Name the raw materials for the Haber process, and state the source of each one.

2 Evaluate the temperature chosen for the Haber process in terms of yield, environmental impact, and energy requirements.

3 Evaluate the choice of pressure for the Haber process in terms of yield and cost.

**H**

## Equilibrium

Many reactions are reversible. This means they can go in both directions. For example:

- Ammonia gas and hydrogen chloride gas react to form ammonium chloride at room temperature:

  ammonia + hydrogen chloride → ammonium chloride

  $NH_3\,(g)\ +\ HCl\,(g)\ \rightarrow\ NH_4Cl\,(g)$

- If you heat ammonium chloride, it decomposes to make ammonia and hydrogen chloride:

  ammonium chloride → ammonia + hydrogen chloride

  $NH_4Cl\,(g)\ \rightarrow\ NH_3\,(g)\ +\ HCl\,(g)$

In a **closed system**, substances cannot enter or leave the reaction vessel. If a reversible reaction happens in a closed system, it reaches **dynamic equilibrium**. In dynamic equilibrium, the forward and backward reactions are both happening. The rates of both reactions are the same. The amount of each substance in the reaction mixture does not change. For example:

ammonia + hydrogen chloride ⇌ ammonium chloride

$NH_3\,(g)\ +\ HCl\,(g)\ \rightleftharpoons\ NH_4Cl\,(g)$

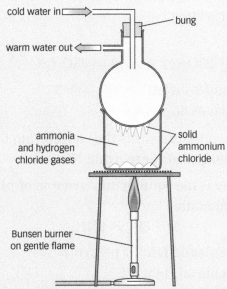

▲ The equilibrium of ammonia, hydrogen chloride, and ammonium chloride.

## Changing equilibrium conditions

The amounts of substances at equilibrium depend on the conditions, including:

- temperature (for all reactions)
- pressure (for gaseous reactions).

For any equilibrium reaction, if conditions are changed the reaction tends to counteract the effect of the change.

### Revision objectives
- explain what equilibrium means
- explain how changing the temperature and pressure affects equilibrium reactions

### Student book references
3.25 Equilibrium reactions
3.26 Heating, cooling, and equilibrium
3.27 Pressure and equilibrium

### Specification key
✔ C3.5.1 c – h

# Key words

**closed system,
dynamic equilibrium**

## Exam tip    AQA

Remember, lowering
temperature increases the yield
for the exothermic process
and decreases the yield for the
endothermic process.

## Questions

1   Explain what an
    equilibrium reaction is.
    Give an example and an
    equation to illustrate your
    answer.

2   Explain the effect of
    changing the pressure on
    an equilibrium reaction in
    the gas phase.

### Working to Grade E

1 Complete the table to show the sources of the raw materials for the Haber process.

| Raw material | Source |
|---|---|
| nitrogen | |
| hydrogen | |

2 Write the conditions that are usually chosen for the Haber process in the table below.

| temperature (°C) | |
|---|---|
| pressure (atmospheres) | |
| catalyst | |

### Working to Grade C

3 The diagram shows how ammonia is made in the Haber process. Write the letters of the labels below in the correct boxes. You may write one or more letters in each box.

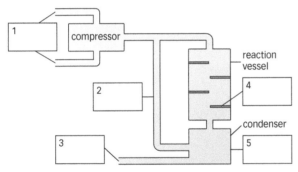

A Purified raw materials enter the apparatus here.
B Unused nitrogen and hydrogen travel through these pipes so that they can be added to the reaction mixture again.
C Liquid ammonia is removed here.
D The reaction mixture is cooled here.
E Iron catalyst.

4 The equation below summarises the processes that occur in the Haber process.

$$N_2(g) + 3H_2(g) \rightleftharpoons 2NH_3(g)$$

Write **T** next to the statements below that are true. Write corrected versions of the statements that are false.

a The symbol $\rightleftharpoons$ shows that the reaction is reversible.
b In the reaction vessel, ammonia molecules break down to make nitrogen and steam.
c In the reaction vessel, nitrogen and hydrogen react together to make ammonia.
d One mole of nitrogen reacts with three moles of hydrogen to make six moles of ammonia.
e There are three gases in the reaction vessel.

### Working to Grade A*

5 The equation below represents the equilibrium between two gases – nitrogen dioxide and dinitrogen tetroxide.

$$N_2O_4(g) \rightleftharpoons 2NO_2(g) \; \Delta H = +58 \text{ KJ/mol}$$

Tick the boxes in the table to show how changing the temperature and pressure affect the position of the equilibrium.

| Change | Effect on position of equilibrium | | |
|---|---|---|---|
| | Shifts left | No change | Shifts right |
| increasing temperature | | | |
| decreasing pressure | | | |
| adding a catalyst | | | |

6 For each change in the table above, explain **why** the position of the equilibrium changes (or does not change) in the way you have indicated.

7 Write **T** next to each statement that is true for a system at equilibrium. Write corrected versions of the statements that are false.
a At equilibrium, the rate of the forward reaction is the same as the rate of the backward reaction.
b Each reactant and product is present in the equilibrium mixture.
c At equilibrium, the amounts of products in the mixture gradually increase.
d At equilibrium, the amounts of the reactants in the mixture gradually decrease.
e Equilibrium can be approached from the product side or the reactant side.
f Equilibrium can only be reached in a closed container.

8 The graph shows the relationship between pressure and yield for the Haber process at 450 °C.

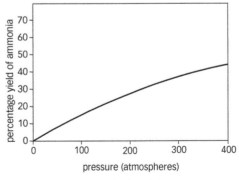

a Describe the relationship shown by the graph.
b Explain why a pressure of 200 atm is chosen for the Haber process.

**1** Ammonia is a vital chemical. It is used to make fertilisers.

The chemical industry uses the Haber process to manufacture huge amounts of ammonia.

**a** The Haber process is based on the reaction below.

$$\text{nitrogen} + \text{hydrogen} \rightleftharpoons \text{ammonia}$$

**i** Explain what the symbol $\rightleftharpoons$ tells you about the reaction.

.........................................................................................................................................................................

*(1 mark)*

**ii** The symbol equation for the reaction is given below. It is not balanced.

Write one number on each dotted line to balance the equation.

$$N_2\,(g) + \text{.........}H_2(g) \rightleftharpoons \text{.........}NH_3\,(g)$$

*(1 mark)*

**b** The diagram below summarises the stages of the Haber process.

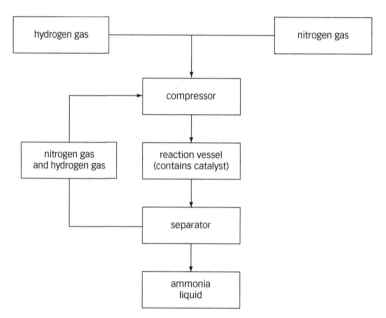

**i** Explain why unreacted nitrogen and hydrogen are recycled.

.........................................................................................................................................................................

.........................................................................................................................................................................

*(2 marks)*

**ii** Give the purpose of the catalyst.

.........................................................................................................................................................................

*(1 mark)*

**iii** In the Haber process, energy is transferred from the cooling gases in the condenser to the reaction vessel.

Give **two** reasons for this.

**1** ..............................................................................................................................................................

**2** ..............................................................................................................................................................

*(2 marks)*

**c** The graph shows the relationship between temperature, pressure, and yield of ammonia.

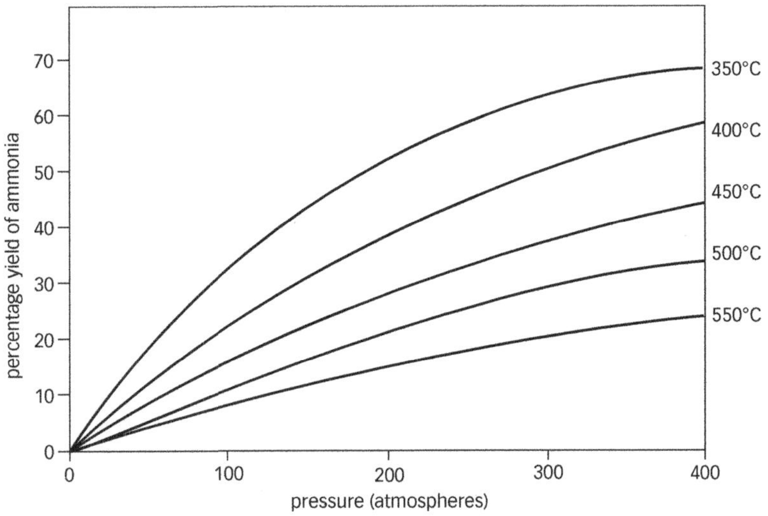

**i** Use the graph to describe the relationship between **temperature** and **yield** of ammonia.

...............................................................................................................................................................

*(1 mark)*

**ii** Use your answer to part **ci**, and your own knowledge, to explain why a temperature of 450 °C is often chosen for the Haber process.

In your answer, refer to economic factors and energy use.

...............................................................................................................................................................

...............................................................................................................................................................

...............................................................................................................................................................

*(3 marks)*

***(Total marks: 11)***

**H**

**2** Every year, chemical plants all over the world produce a total of around 150 million tonnes of sulfuric acid.

The equilibrium reaction below is an important stage in the process of the manufacture of sulfuric acid.

sulfur dioxide + oxygen ⇌ sulfur trioxide

**a** Which of the following statements about the equilibrium reaction are true?

Tick boxes to show the best answers.

In the equilibrium mixture, sulfur dioxide and oxygen are reacting to make sulfur trioxide. ☐

The amount of sulfur dioxide in the equilibrium mixture decreases all the time. ☐

The forward reaction is faster than the backward reaction. ☐

The amount of sulfur trioxide in the equilibrium mixture increases all the time. ☐

In the equilibrium mixture, sulfur trioxide is decomposing to make sulfur dioxide and oxygen. ☐

The backward reaction is faster than the forward reaction. ☐

*(2 marks)*

**b** Look at the balanced equation below.

$$2SO_2(g) + O_2(g) \rightleftharpoons 2SO_3 \; \Delta H = -197 \, KJ/mol$$

Use the equation and the energy change for the reaction to predict and explain the effects on the yield of sulfur trioxide of changing the temperature and pressure.

*In this question you will get marks for using good English, organising information clearly, and using scientific words correctly.*

..........................................................................................................................................

..........................................................................................................................................

..........................................................................................................................................

..........................................................................................................................................

..........................................................................................................................................

..........................................................................................................................................

..........................................................................................................................................

..........................................................................................................................................

*(6 marks)*

*(Total marks: 8)*

## What's in an alcohol?

Alcohols are **organic compounds**. This means that their molecules are made up mainly of carbon and hydrogen atoms.

There are many **alcohols**. They all include the reactive –OH group. A reactive group of atoms in an organic molecule is called a **functional group**.

The alcohols in the table below are members of the same **homologous series**. The members of a homologous series have the same functional group, but different numbers of carbon atoms.

| Name | Molecular formula | Structural formula |
|------|-------------------|--------------------|
| methanol | $CH_3OH$ | 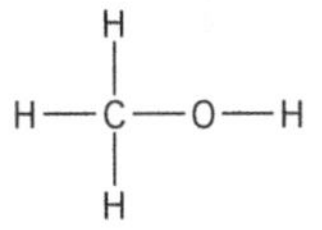 |
| ethanol | $CH_3CH_2OH$ | 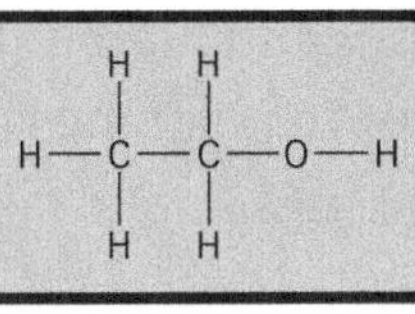 |
| propanol | $CH_3CH_2CH_2OH$ | 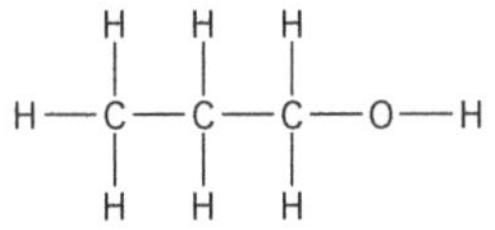 |

## Properties of alcohols

Methanol, ethanol, and propanol are in the same homologous series. They have similar properties.

- They dissolve in water to form neutral solutions (pH 7).
- They react with sodium. The products are a salt and hydrogen. For example:

   methanol + sodium → sodium methoxide + hydrogen

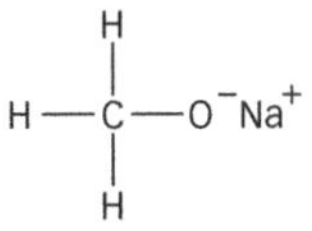

   ethanol + sodium → sodium ethoxide + hydrogen

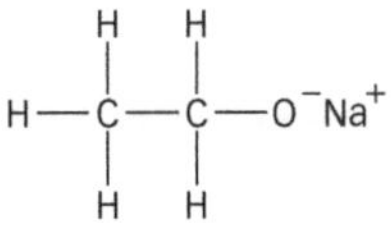

- They burn in air. The products are carbon dioxide and water. For example:

$$\text{ethanol} + \text{oxygen} \rightarrow \text{carbon dioxide} + \text{water}$$
$$CH_3CH_2OH + 3O_2 \rightarrow 2CO_2 + 3H_2O$$
$$\text{propanol} + \text{oxygen} \rightarrow \text{carbon dioxide} + \text{water}$$
$$2CH_3CH_2CH_2OH + 9O_2 \rightarrow 6CO_2 + 8H_2O$$

### Revision objectives

- recognise alcohols from their names or formulae
- describe the reactions of alcohols
- evaluate the social and economic advantages and disadvantages of the uses of alcohols

### Student book references

**3.28** Alcohols – 1

**3.29** Alcohols – 2

### Specification key

- C3.6.1

# Using alcohols

## Alcoholic drinks

Ethanol is the alcohol in alcoholic drinks. Alcoholic drinks have social and economic advantages and disadvantages.

|          | Advantages | Disadvantages |
|----------|-----------|---------------|
| Social   | • Makes people feel relaxed for a short time. | • Slow reaction time – increases risk of road accidents.<br>• Make people forgetful, confused, and more likely to act foolishly.<br>• Cause vomiting, unconsciousness, death. |
| Economic | • Profitable for drinks companies.<br>• Taxes provide income for government. | • Treating alcohol-related health problems is costly.<br>• Dealing with alcohol-related crime is costly. |

## Alcohols as solvents

Ethanol and methanol are useful solvents. Ethanol is used as a solvent in perfumes and deodorants, medicines, and food flavourings.

## Alcohols as fuels

When alcohols burn in air, much energy is released as heat. So alcohols are useful fuels. There are advantages and disadvantages of using ethanol as a fuel instead of fossil fuels, such as petrol and diesel.

|               | Advantages | Disadvantages |
|---------------|-----------|---------------|
| Environmental | • The crops from which ethanol is manufactured take in carbon dioxide gas as they grow. Some people say this means that ethanol fuel is **carbon neutral**. | • On burning, alcohols produce carbon dioxide gas.<br>• Making fertilisers for the crops produces carbon dioxide. Some people say this means that ethanol fuels are not carbon neutral. |
| Social        | • Made from renewable resources such as sugar cane or maize. | • Crops from which ethanol is made are grown on land that could be used to grow food. |

# Oxidation reaction

Ethanol can be oxidised to ethanoic acid. It is oxidised by:
- chemical oxidising agents, for example, potassium dichromate(VI) solution
- the action of microbes.

Ethanoic acid is the main acid in vinegar.

## Exam tip

You need to be able to write balanced chemical equations for the *combustion* reactions of alcohols, but not for any of their other reactions.

**Questions**

1   Write the functional group that is in all alcohols.

2   Draw a table to summarise the chemical reactions of ethanol.

3   Write a balanced symbol equation for the combustion reaction of methanol.

## What's in a carboxylic acid?

Methanoic acid, ethanoic acid, and propanoic acid are members of the same homologous series. They are all **carboxylic acids**.

Every carboxylic acid molecule includes the functional group –COOH. The atoms in the functional group are arranged like this:

The table shows the formulae of three carboxylic acids.

| Name | Molecular formula | Structural formula |
|---|---|---|
| methanoic acid | HCOOH | |
| ethanoic acid | $CH_3COOH$ | |
| propanoic acid | $CH_3CH_2COOH$ | |

## Properties of carboxylic acids

Methanoic acid, ethanoic acid, and propanoic acid have similar properties.

• They dissolve in water to form acidic solutions (with pH less than 7).
• They react with carbonates. The products are a salt, carbon dioxide, and water. For example:

ethanoic + sodium → sodium + carbon + water
acid        carbonate   ethanoate   dioxide

• They react with alcohols to make esters.

# Using carboxylic acids

Ethanoic acid is the main acid in vinegar. Citric acid is added to some foods and drinks to give them a sour taste.

Many fruits and vegetables contain ascorbic acid, vitamin C. This is vital for health. Vitamin tablet companies include ascorbic acid in their tablets.

The medicine aspirin is also a carboxylic acid. It is a painkiller. Aspirin also reduces blood clotting, so is taken by some people at risk of heart attacks.

## H Weak acids

When ethanoic acid dissolves in water, some of its molecules split up. Two ions are formed:
- a positive hydrogen ion
- a negative ethanoate ion.

The ethanoic acid molecule has **ionised**.

Ethanoic acid is a **weak acid** because fewer than 1% of its molecules ionise when they dissolve in water. The equilibrium for the solution of ethanoic acid lies to the left:

$$CH_3COOH(l) + (aq) \rightleftharpoons CH_3COO^-(aq) + H^+(aq)$$

All carboxylic acids are weak acids.

Hydrochloric acid also ionises when it dissolves. All its molecules split up to form hydrogen ions and chloride ions:

$$HCl(g) + (aq) \rightarrow H^+(aq) + Cl^-(aq)$$

Hydrochloric acid is a **strong acid**, because all its molecules ionise when it dissolves.

## Weak acids and pH

pH values measure the acidity of a solution. The lower the pH, the more acidic the solution. The pH of a solution tells us about the concentration of hydrogen ions – the greater the concentration of hydrogen ions, the lower the pH.

The table gives the pH of two acids of the same concentration.

| | |
|---|---|
| hydrochloric acid, 0.1 mol/dm³ | pH 1.0 |
| ethanoic acid, 0.1 mol/dm³ | pH 2.9 |

The ethanoic acid has a smaller concentration of hydrogen ions in solution. This is because only 1% of the ethanoic acid molecules are ionised. The ethanoic acid has a higher pH.

## Questions

1 Draw a table to summarise the reactions of the carboxylic acids.

2 List **four** uses of carboxylic acids.

3 H Predict and explain the difference in pH of a 0.5 mol/dm³ solution of hydrochloric acid and a solution of ethanoic acid of the same concentration.

## What's in an ester?

Esters include the functional group –COO– in their molecules. The atoms in the functional group are arranged like this:

$$-C-O-$$
$$\parallel$$
$$O$$

There are many esters. One of them is ethyl ethanoate. Its structural formula is below.

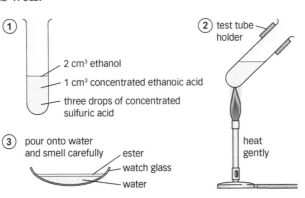
ethyl ethanoate

## Properties of esters

Esters share many properties. They:
- are **volatile,** meaning that they easily form vapours
- have distinctive smells – for example, the ester pentyl ethanoate contributes to the smell of pears.

## Using esters

The smells and tastes of esters make them useful for:
- making perfumes, shampoos, and shower gels
- flavouring foods such as sweets and chocolates.

## Making esters

You can make esters in the laboratory by reacting a carboxylic acid with an alcohol. The process is called **esterification**. For example, ethanoic acid reacts with ethanol to make ethyl ethanoate. You need to use an acid catalyst – concentrated sulfuric acid works well.

The diagrams show what to do.

① 2 cm³ ethanol
1 cm³ concentrated ethanoic acid
three drops of concentrated sulfuric acid

② test tube holder
heat gently

③ pour onto water and smell carefully
ester
watch glass
water

The equation below summarises the reaction in which ethyl ethanoate is made from ethanol and ethanoic acid.

ethanoic acid  +  ethanol  →  ethyl ethanoate  +  water

### Questions

1. Draw the structural formula of ethyl ethanoate.

2. Write a word equation for the reaction of ethanol with ethanoic acid to form an ester.

3. Write down **two** uses of esters.

## Working to Grade E

**1** Draw lines to match each compound to one use.

| Compound |
|---|
| ethanol |
| ethanoic acid |
| pentyl pentanoate |

| Use |
|---|
| alcoholic drinks |
| flavouring |
| vinegar |

**2** Highlight the one correct word in each pair of **bold** words.

Alcohols dissolve in water to form **neutral/acidic** solutions. Alcohols react with sodium to produce **hydrogen/water**. Carboxylic acids dissolve in water to form **neutral/acidic** solutions. Carboxylic acids react with carbonates to produce **hydrogen/carbon dioxide**.

**3** Complete the table below.

| Name of group of organic compounds | Examples |
|---|---|
| alcohols | • ethanol<br>• |
| carboxylic acids | •<br>• |
|  | • ethyl ethanoate<br>• |

**4** Use the words in the box below to complete the sentences that follow. Each word may be used once, more than once, or not at all.

> homologous  propanoic  microbes  ethanoic
> solvents  fuels  perfumes  functional  propanol
> ethanol  oxidising  vinegar  flavourings  drinks

Alcohol molecules contain the _____ group –OH. Methanol, ethanol, and _____ are members of the same _____ series. Alcohols are used as _____, _____ and in _____. The main alcohol in alcoholic drinks is _____. Ethanol can be oxidised to _____ acid either by chemical _____ agents or by the action of _____.

## Working to Grade C

**5** Write **T** next to the sentences below that are true. Write corrected versions of the sentences that are false.
  **a** Carboxylic acids have the functional group –COOOH
  **b** Carboxylic acids react with acids to produce esters.
  **c** The molecular formula of methanoic acid is $CH_3COOH$.
  **d** Ethanoic acid is the main acid in vinegar.
  **e** Vinegar is acidic.
  **f** Propyl propanoate is a carboxylic acid.

**6** Fill in the empty boxes.

| Name | Molecular formula | Structural formula |
|---|---|---|
| methanol | | |
| | $CH_3CH_2OH$ | |
| | | |
| | HCOOH | |
| | | |
| propanoic acid | | |

**7** For each list below, circle **two** compounds that react together to make an ester.
  **a** Ethanol, propane, propanoic acid, ethyl propanoate.
  **b** Propane, propanol, ethanoic acid, ethane.
  **c** Water, methanoic acid, methane, methanol.
  **d** Ethanoic acid, ethyl ethanoate, ethanol, ethane.

**8** Complete the word equations below.
  **a** ethanol + oxygen → _____ + water
  **b** methanol + sodium → sodium methoxide + _____
  **c** ethanol + sodium → _____ + _____
  **d** propanol + oxygen → _____ + _____
  **e** ethanoic acid + sodium carbonate → sodium ethanoate + _____ + water
  **f** propanoic acid + calcium carbonate → _____ + carbon dioxide + _____
  **g** _____ + ethanoic acid → ethyl ethanoate + water
  **h** propanol + propanoic acid → _____ + water

## Working to Grade A*

**9** Balance the equations below.
  **a** $CH_3OH + O_2 \rightarrow CO_2 + H_2O$
  **b** $CH_3CH_2OH + O_2 \rightarrow CO_2 + H_2O$

**10** Explain the meaning of the statement below:
  *Ethanoic acid is a weak acid.*

**11** Matthew measures the pH of three acids, A, B, and C. Each acid has the same concentration. His results are in the table.
Matthew knows that one of the acids is hydrochloric acid, and that the other two are carboxylic acids.

| Acid | pH |
|---|---|
| A | 1.2 |
| B | 3.7 |
| C | 4.1 |

  **a** Which acid is hydrochloric acid? Explain how you decided.
  **b** Is acid C a stronger or weaker acid than acid B? Explain how you decided.

# Examination questions
## Alcohols, carboxylic acids, and esters

**1** The solvent in this nail varnish is ethyl ethanoate.

**a** Which formula represents the structure of ethyl ethanoate?

Draw a ring around the correct formula.

*(1 mark)*

**b** To make ethyl ethanoate, scientists react two substances together.

Tick the names of the **two** chemical families to which these substances belong.

alcohols

esters

carboxylic acids

alkanes

*(1 mark)*

**c** Sulfuric acid is added to the reaction mixture during the synthesis of ethyl ethanoate. Explain why.

......................................................................................................................................................................

*(1 mark)*

**d** Pentyl pentanoate has the same functional group as ethyl ethanoate.

It is used as a food flavouring.

Write down the property of pentyl pentanoate that makes it suitable for use as a food flavouring.

......................................................................................................................................................................

*(1 mark)*

**(Total marks: 4)**

2 Identify **three** uses of alcohols.

Choose **one** of these uses, and evaluate the social and economic advantages and disadvantages of this use.

*In this question you will get marks for using good English, organising information clearly, and using scientific words correctly.*

........................................................................................................................................................

........................................................................................................................................................

........................................................................................................................................................

........................................................................................................................................................

........................................................................................................................................................

........................................................................................................................................................

........................................................................................................................................................

........................................................................................................................................................

........................................................................................................................................................

........................................................................................................................................................

*(6 marks)*

*(Total marks: 6)*

3 This question is about hexanoic acid.

**a** Complete the diagram below to show the structural formula of hexanoic acid.

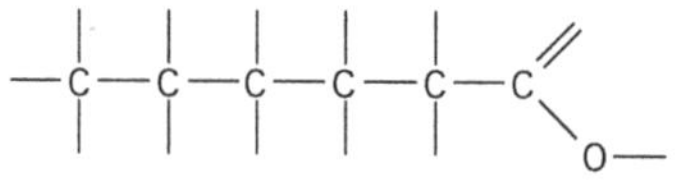

*(1 mark)*

**b** Hexanoic acid reacts with calcium carbonate.

Complete the word equation for the reaction below.

hexanoic acid + calcium carbonate → calcium hexanoate + ............................ + water

*(1 mark)*

c   A student tests the pH of four organic compounds.

Her results are in the table.

| Compound | pH |
|---|---|
| A | 4.3 |
| B | 7.0 |
| C | 10.0 |
| D | 12.2 |

Which of the compounds in the table could be hexanoic acid?

..........................................................................................................................................................

*(1 mark)*

**H**

d   i   Explain what makes hexanoic acid a weak acid.

..........................................................................................................................................................

..........................................................................................................................................................

*(2 marks)*

ii   A student has samples of the acids listed below.

Predict which of the acids has the highest pH value.

Explain your choice.

**List of acids**
0.1 mol/dm³ hydrochloric acid
0.1 mol/dm³ sulfuric acid
0.1 mol/dm³ nitric acid
0.1 mol/dm³ ethanoic acid

..........................................................................................................................................................

..........................................................................................................................................................

*(1 mark)*

*(Total marks: 6)*

## Designing an investigation and making measurements

In this module there are several opportunities to design investigations and make measurements. These include investigating the volumes of acids and alkalis that react together, and measuring the pH of solutions of alcohols and carboxylic acids.

As well as demonstrating your investigative skills practically, you are likely to be asked to comment on investigations done by others. The examples below offer guidance in these skill areas. They also give you the chance to practise using your skills to answer the sorts of question that may come up in exams.

## The volumes of acid and alkali solutions that react with each other

Suzanna tested the hypothesis that the higher the concentration of hydrochloric acid, the smaller the volume required to neutralise an alkali. She did titrations to measure the volumes of hydrochloric acid of different concentrations that reacted with sodium hydroxide solution. She used the apparatus below.

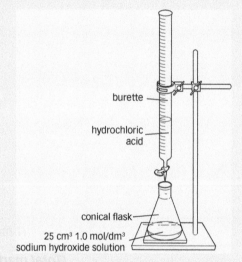

1 Suzanna did a trial run before she started her main investigation. Suggest **three** questions Suzanna might have wanted to consider in her trial run.

In general, a trial run helps an investigator to select appropriate values to be recorded.

In your answer to this question, suggest specific questions Suzanna might have wanted to address before starting her main investigation, such as 'what volume of acid shall I use?'.

2 Suzanna's results are in the table.

| Concentration of hydrochloric acid (mol/dm³) | Volume of acid required to neutralise 25 cm³ of 1.0 mol/dm³ sodium hydroxide solution (cm³) | | | |
|---|---|---|---|---|
| | Run 1 | Run 2 | Run 3 | Mean |
| 0.5 | 49.50 | 49.60 | 49.40 | 49.50 |
| 0.8 | 31.05 | 31.00 | 30.95 | 31.00 |
| 1.0 | 25.15 | 25.10 | 25.05 | 25.10 |
| 1.5 | 17.20 | 16.60 | 16.00 | 16.60 |
| 2.0 | 12.50 | 12.50 | 12.50 | 12.50 |

Identify the independent and dependent variables.

- The independent variable is the one that is changed or selected by the investigator.
- The dependent variable is measured for each change in the independent variable.

3 Describe how Suzanna can make sure her measurements are valid.

For a measurement to be valid it must measure only the appropriate variable – how can Suzanna ensure she does this? One answer is to keep the volume of acid the same in each run. Can you think of others?

4 Describe how Suzanna can make sure she does a fair test.

In a fair test, only the independent variable should affect the dependent variable. All the other variables must be kept the same.

In your answer, include a list of the control variables.

5 Compare the precision of the data obtained when the concentration of hydrochloric acid was 1.0 mol/dm³ and when the concentration of hydrochloric acid was 1.5 mol/dm³.

Data is precise if values obtained by repeated measurements cluster closely. In Suzanna's investigation:

- If the values obtained for a certain acid concentration are close together, the measurements for this concentration are precise.
- If the values for a certain acid concentration are not close together, the measurements for this concentration are not precise.

6 Suggest why the values obtained for repeated readings are not exactly the same.

There will always be some variation in the actual value of a variable, no matter how hard an investigator tries to repeat an event. In answering this question, think about what Suzanna might have done to have obtained results that are slightly different, or how the equipment she used might have led to slightly different results for any given acid concentration.

# Answering a question with data response

*In this question you will be assessed on using good English, organising information clearly, and using specialist terms where appropriate.*

**1** A student wants to identify two salts. The table below shows the tests he did, and gives his results. Identify salts A and B, using data from the table to explain and support your decisions. *(6 marks)*

QUESTION

| Test number | Test | Salt A observations | Salt B observations |
|---|---|---|---|
| 1 | Add sodium hydroxide solution. | blue precipitate | green precipitate |
| 2 | Add dilute hydrochloric acid to a sample of the solid salt. | fizzing – gas produced that made limewater cloudy | no change |
| 3 | Dissolve a little of the salt in water. Add barium chloride solution and hydrochloric acid. | no change | precipitate formed on filtering, could see that precipitate was white |
| 4 | Dissolve a little of the salt in water. Add silver nitrate solution and nitric acid. | no change | no change |

**G–E**

B is iron (III) chloride – this is wot the test reasults show me. And A contains copper. And B has chloride ions in it. But I dont no wot the other bit of A is. I think he needs to do some more tests if he has time he could try tests with pottasium and sodium these are fun and can eksplode.

**Examiner:** This answer is typical of a grade-G candidate. It is worth just one mark, gained for recognising that salt A contains copper ions.

There are several spelling mistakes, and the answer is not well organised. The last sentence is not relevant.

**D–C**

A is copper (II) carbonate. I know it has carbonate because the solid fizzed when he drip acid onto it and the gas made limewater milky because it is carbon dioksyd.

B is definitely iron (II) sulfate. I be sure about this becaus of the test reasults.

**Examiner:** This answer is worth three marks out of six. It is typical of a grade-C or -D candidate.

The candidate has correctly identified both salts, but has only given reasons to support the identification of one ion (carbonate).

The answer is well organised, with a few mistakes of grammar and spelling.

**B–A\***

Salt A is copper (II) carbonate. The test with sodium hydroxide shows that the salt contains copper (II) ions. Test number two shows that salt A is a carbonate, since when it reacted with acid it made a gas that made limewater cloudy (carbon dioxide). Test 1 shows that salt B contains iron (II) ions. The green precipitate is iron (II) hydroxide. Test 3 shows that salt B also contains sulfate ions. The precipitate is barium sulfate.

The student had to filter the mixture for the barium chloride test, otherwise he would not have known that the precipitate was white, since the solution of the salt in water must have been coloured. Together, tests 1 and 3 show that salt B is iron (II) sulfate.

**Examiner:** This is a high-quality answer, typical of an A\* candidate. It is worth six marks out of six.

The candidate has correctly identified the two salts, and given clear and detailed reasons to support his decisions.

The answer is well-organised. The spelling, punctuation, and grammar are faultless. The candidate has used several specialist terms.

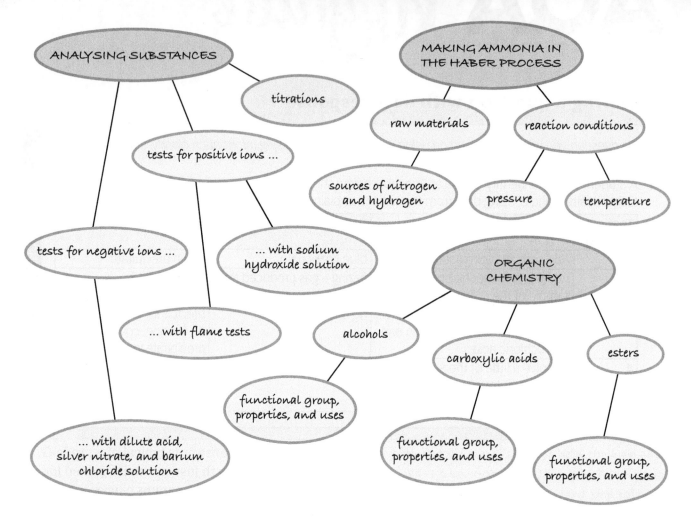

## Revision checklist

○ Many metal compounds give distinctive flame colours, including compounds of lithium (crimson), sodium (yellow), potassium (lilac), calcium (red), and barium (green).

○ You can identify some metal ions in solution by adding sodium hydroxide solution. The precipitates formed have distinctive colours.

○ You can test for carbonates by adding dilute acid, for halide ions by adding silver nitrate solution, and for sulfates by adding barium chloride solution.

○ You can use titrations to measure the volumes of acid and alkaline solutions that react with each other.

○ If you know the concentration of one reactant, you can use titration results to calculate the concentration of the other reactant.

○ The raw materials for the Haber process are nitrogen (obtained from the air) and hydrogen (often obtained from natural gas).

○ In the Haber process, nitrogen and hydrogen react together to form ammonia, $NH_3$, in a reversible reaction. The reaction takes place at high temperature (450 °C), high pressure (200 °C), and in the presence of an iron catalyst.

○ When a reversible reaction occurs in a closed system, equilibrium is reached when the reactions occur at the same rate in each direction.

○ The relative amounts of the reacting substances at equilibrium depend on the conditions of the reaction. Changing the temperature or pressure changes the amounts of substances at equilibrium.

○ Alcohols have the functional group –OH. They form neutral solutions in water, react with sodium to produce hydrogen, and burn in air. Ethanol is oxidised to ethanoic acid.

○ Alcohols are used as fuels, solvents, and in drinks.

○ Carboxylic acids have the functional group –COOH. They form acidic solutions in water, react with carbonates to produce carbon dioxide, and react with alcohols to produce esters.

○ Ethanoic acid is the main acid in vinegar.

○ Carboxylic acids are weak acids because they do not ionise completely in water.

○ Esters have the functional group –COO–. They are volatile compounds with distinctive smells.

○ Esters are used as flavourings and perfumes.

# Electromagnetic waves

There are many different types of **electromagnetic wave**. Although they all have the same speed in a vacuum (the speed of light), the energy that they carry depends on their **wavelength**.

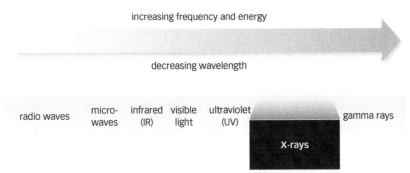

▲ X-rays are at the high-energy end of the electromagnetic spectrum.

**X-rays** are produced in hospitals from fast-moving electrons and have a very short wavelength (about the same size as an atom). Only gamma rays have a shorter wavelength. X-rays share two important properties with gamma rays:
- they are very penetrating
- they are ionising.

Both of these properties make X-rays useful for medicine.

# Penetration

As X-rays pass through matter, they are **absorbed**, knocking electrons off atoms (**ionisation**). High-density materials, such as metals and bone, are better at doing this than low-density materials, such as skin and muscle.

# Ionisation

Human cells are damaged when they absorb X-rays. The ionisation of the cell can alter its DNA, making it into a cancer. People who work with X-rays must therefore take these precautions:
- stand behind lead or concrete shielding
- leave the room while the X-ray machine is on
- wear a monitoring badge
- limit the amount of time that they spend working with X-rays.

The damaging effect of X-rays is sometimes put to use in **radiotherapy**. This uses thin beams of X-rays to destroy cancer tumours deep inside the body, without the need for surgery.

## Revision objectives

✔ place X-rays correctly in the electromagnetic spectrum

✔ explain the use of X-rays to make images and diagnoses in medicine

✔ explain the safety precautions required for X-rays

## Student book references

**3.1** X-rays

**3.2** Using X-rays

## Specification key

✔ P3.1.1

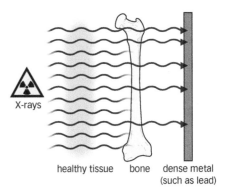

▲ X-rays are easily transmitted through healthy tissue; bone absorbs some X-rays, and lead absorbs most X-rays.

electromagnetic wave,
wavelength, X-ray, absorbed,
ionisation, radiotherapy,
photographic film, CT scan,
charge-coupled device (CCD)

## Photos

Dentists routinely use **photographic film** to make X-ray images of people's teeth. The film needs to be wrapped in black plastic because it is sensitive to light. The film is placed at one side of the teeth and the X-ray source at the other side. When the source is switched on, the intensity of X-rays reaching each part of the film depends on the density of the material in the way:

- Metal fillings completely absorb X-rays.
- Tooth and bone partially absorb X-rays.
- Skin, muscle, and gum hardly absorb X-rays at all.

The film needs to be processed chemically to fix the image. White regions show where the X-rays didn't reach the film. The rest of the film turns black or grey, depending on the dose of X-rays that reached it. The image can be used to show the dentist any decay inside a tooth.

## CT scans

X-rays can be used to make images of more than just bones. A computed tomography (**CT**) **scan** can create images of internal organs, such as livers and kidneys, on a computer screen. A beam of X-rays passes through the patient and the intensity recorded by a **charge-coupled device (CCD)**. The beam and detector are slowly rotated around the patient so that X-rays eventually pass through in all directions. The computer combines all of the information to generate an image of the slice of patient that the X-rays passed through.

### Exam tip

Don't confuse X-rays with X-ray images.

### Questions

1 What is the advantage of imaging a broken bone with X-rays?

2 Explain how an image of a broken bone can be created using X-rays.

3 **H** Explain the use of X-rays to diagnose and treat a cancer tumour deep inside the body.

## Sound

A **sound wave** is created whenever part of a material is made to vibrate. The wave carries the vibrations through the material. The material can be solid, liquid, or gas. The **frequency** of the wave in hertz (Hz) is the number of complete vibrations of the source in one second. Humans can hear sounds in air with frequencies between 20 Hz and 20 000 Hz (20 kHz). Sound that has a frequency above this range is called **ultrasound** – it is widely used to image internal organs in the human body.

## Vibrating crystals

Pulses of ultrasound for medical scans are produced by making special crystals vibrate at the required high frequency. Regular electrical pulses across the **crystal** generate short bursts of ultrasound that move away from the crystal through the medium. Any sudden change in density of the medium causes some of the ultrasound to be reflected back to the crystal. As the reflections are absorbed by the crystal it makes electrical signals. So the crystal acts as both a **transmitter** and **receiver** of ultrasound.

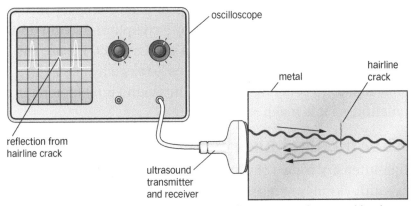

▲ Any change of density of the medium reflects some ultrasound back to its source.

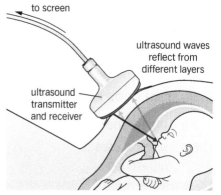

▲ Using ultrasound to image an unborn baby.

# Scanning

Ultrasound is routinely used to scan pregnant women to make images of their unborn babies. A pulsed ultrasound transmitter is held against the skin above the baby. As each pulse passes a boundary inside the woman's body, some of it is reflected back to the transmitter. The depth of the boundary is half the total distance travelled by the pulse. You can use this equation to calculate the total distance travelled by the ultrasound.

$$d = v \times t$$

$d$ is the total distance travelled by the ultrasound in metres (m)
$v$ is the speed of the ultrasound in metres per second (m/s)
$t$ is the time delay for the reflected pulse in seconds (s)

The average speed of sound in the human body is 1500 m/s. So if the time between sending out a pulse of ultrasound and receiving a reflection is 60 microseconds, how far below the skin is the reflecting boundary?

$$d = v \times t = 1500\,\text{m/s} \times 60 \times 10^{-6}\,\text{s} = 9.0 \times 10^{-2}\,\text{m}$$

The wave travels $9.0 \times 10^{-2}$ m in total, so the reflecting boundary must be half of this distance ($4.50 \times 10^{-2}$ m) below the skin.

# Safe images

For each pulse fired into the mother from the transmitter, there are many reflected pulses, one for each boundary inside her. This information can be used by a computer to build up an image of the baby. The technique is similar to a CT scan, but without using X-rays. This is good, as X-rays may harm the unborn baby by ionising its cells. Ultrasound is a non-ionising radiation, so it is much safer.

# Other medical uses

High-intensity beams of ultrasound can also be used to break up painful kidney stones into small pieces so that they flush out with the patient's urine.

## Questions

1  What is ultrasound?

2  Explain how ultrasound can be used to make an image of an unborn baby.

3  **H** A pulse of ultrasound is reflected off a bone 10 cm below the skin. If ultrasound travels at a speed of 1.5 km/s, what is the time delay between transmission and reception of the pulse?

# Questions
# X-rays and ultrasound

## Working to Grade E

1   Which of the following objects has about the same size as the wavelength of X-rays?
         atom   electron   microbe   nucleus

2   Why are unborn babies scanned by ultrasound instead of X-rays?

3   Suggest **three** safety precautions for people who work with X-rays.

4   Which of the following frequencies could be for ultrasound?
         1 Hz   10 Hz   10 kHz   100 kHz

5   What are the differences between a CT scan image and an X-ray image on photographic film?

## Working to Grade C

6   Explain how X-ray images of broken bones can be made on photographic film.

7   Explain **three** safety precautions required for people who work with X-rays.

8   What is ultrasound?

9   Describe how a CT scan is obtained.

10  Explain how ultrasound can be used to image an unborn baby.

## Working to Grade A*

11  The reflection of an ultrasound pulse sent into a material is received after a delay of 40 microseconds. If the speed of ultrasound in the material is 2.5 km/s, how deep in the material is the reflecting boundary?

**1 a** An X-ray image is going to be made on photographic paper of a broken arm.

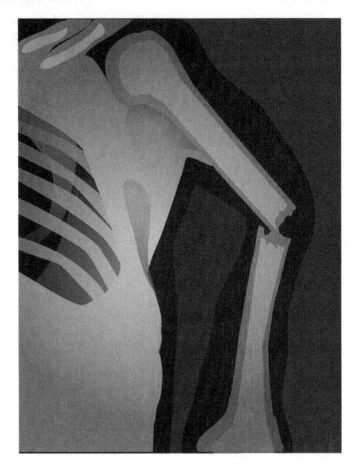

The following sentences explain the process of making the image. The sentences are in the wrong order.

**Q** X-rays pass into the arm.
**R** The X-ray source is switched on.
**S** X-rays ionise parts of the photographic film.
**T** The photographic film is wrapped in black plastic.
**U** The film is developed in chemicals to fix the image.
**V** The broken arm is placed on top of the photographic film.

Arrange these sentences in the correct order. Start with the letter **T**.

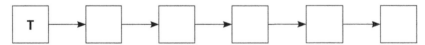

*(4 marks)*

**b** Explain why the radiographer leaves the room while the X-ray source is switched on.

..................................................................................................................................................

..................................................................................................................................................

..................................................................................................................................................

..................................................................................................................................................

*(3 marks)*
*(Total marks: 7)*

**2** The picture shows a pregnant woman having her unborn baby scanned by ultrasound.

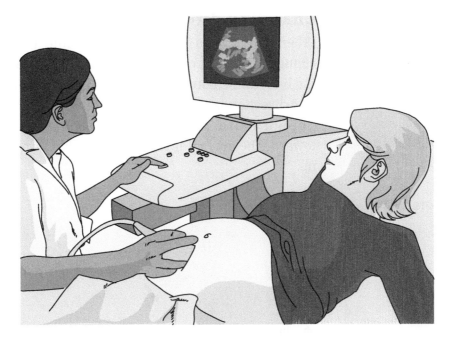

**a** What is ultrasound?

...................................................................................................................................................................

...................................................................................................................................................................

...................................................................................................................................................................

*(2 marks)*

**b** Explain why ultrasound is used for the scan instead of X-rays.

...................................................................................................................................................................

...................................................................................................................................................................

...................................................................................................................................................................

...................................................................................................................................................................

*(3 marks)*

**c** Explain how the image can be obtained by using ultrasound.

...................................................................................................................................................................

...................................................................................................................................................................

...................................................................................................................................................................

...................................................................................................................................................................

...................................................................................................................................................................

...................................................................................................................................................................

*(4 marks)*
***(Total marks: 9)***

## Revision objectives

- ✓ describe the process of refraction
- ✓ use Snell's law
- ✓ understand the term 'refractive index'

## Student book references

**3.5** More on refraction

## Specification key

✓ P3.1.3 a and c

## Key words

refraction, normal, refractive index

## Questions

1 Draw a diagram to show what happens to light when it passes from air into glass.

2 Explain why light refracts when it changes medium.

3 **H** Light is incident at 45° to a plastic block of refractive index 1.6. Calculate the angle of refraction.

# Changing direction

When light travels from one medium into another its speed changes. This can cause **refraction**, a change of direction of the light at the boundary between the two different mediums. The direction is measured as an angle to a construction line (the **normal**) at right angles to the boundary. The light refracts towards the normal if it slows down, but away when it speeds up.

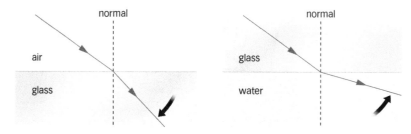

▲ The light refracts towards or away from the normal.

# Refractive index

For light passing into or out of air, the **refractive index** (how fast light travels through the material) of the other medium determines the amount of refraction. The speed of light in a material is related to its refractive index. This equation, called Snell's law, can be used to calculate angles:

$$\text{refractive index} = \frac{\sin i}{\sin r}$$

$i$ is the angle of incidence of the light
$r$ is the angle of refraction of the light

So if light is incident on a plastic block at 45° and refracted at 30°, what is the refractive index of the plastic?

$$\text{refractive index} = \frac{\sin i}{\sin r} = \frac{\sin 45}{\sin 30} = 1.4$$

▲ Increasing the refractive index increases the change of direction.

## Exam tip

Remember that $i$ and $r$ are always measured from the ray to the normal.

# Using refraction

All lenses use refraction of light to make **images**. The refraction happens in two stages: when the light enters the lens and when it leaves. The light bends towards the normal on the way in and away from the normal on the way out.

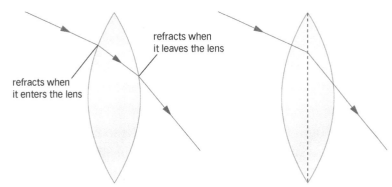

refracts when
it leaves the lens

refracts when
it enters the lens

▲ Ray diagrams of lenses often show the refraction taking place halfway through the lens. This makes the diagrams simpler to draw.

# Converging lens

A **converging** lens takes rays that are parallel to the **principal axis** of the lens and refracts them so that they all pass through the **principal focus**. The distance from the centre of the lens to the principal focus is called the **focal length**.

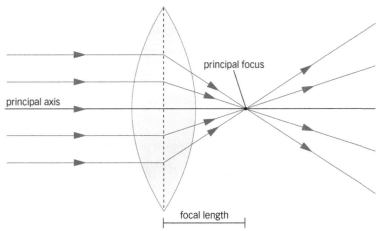

principal focus

principal axis

focal length

▲ Action of a converging lens.

# Diverging lens

A **diverging** lens takes rays that are parallel to the principal axis and refracts them so that they appear to come from the **virtual focus**. This means that the rays spread out after passing through the lens instead of coming together.

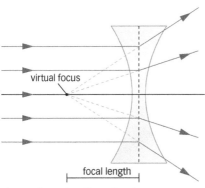

virtual focus

focal length

▲ Action of a diverging lens.

## Revision objectives

- ✔ recognise the difference between converging and diverging lenses
- ✔ use refraction to explain the operation of a lens
- ✔ describe the nature of an image formed by a lens

## Student book references

3.6   Introduction to lenses

3.7   Describing images

## Specification key

✔ P3.1.3 b and d

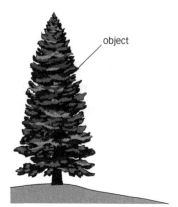

object

convex lens

image

screen

▲ The image of the tree is real, diminished, and inverted.

## Images

A lens takes light from an object and uses it to make an image. The image is a copy of the object, but may be different from it in a number of ways. A number of different terms describe these differences:

- **inverted** means upside down compared with the object
- **upright** means the same way up as the object
- **magnified** means bigger than the object
- **diminished** means smaller than the object
- **real** means that the image can be seen on a screen
- **virtual** means that the image can't be seen on a screen

The image will be formed at a particular distance away from the centre of the lens.

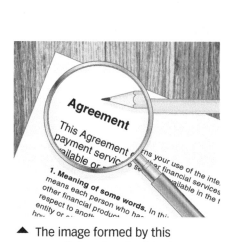

▲ The image formed by this magnifying glass is virtual, magnified, and upright.

## Questions

1  Draw pictures of converging and diverging lenses.

2  Draw a labelled ray diagram for light passing through a converging lens.

3  H Describe the image formed in this picture.

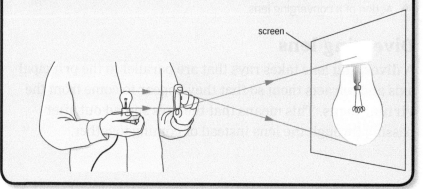

screen

## Drawing rays

A **ray diagram** is a scale drawing that can be used to determine the properties of an object formed by a lens. You draw it on squared paper as follows:

- draw the lens at right angles to the principal **axis**
- draw the **object** as an upright arrow on the principal axis
- draw rays that leave the tip of the object and pass through the lens
- the tip of the **image** is where the rays are focussed by the lens.

Each ray should have an arrow to show its direction. You will need to select suitable scales for the diagram. The vertical and horizontal scales do not have to be the same.

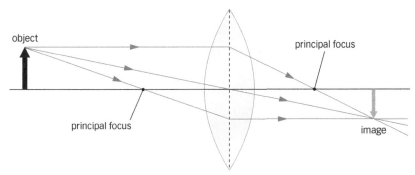

▲ A ray diagram for a converging lens.

## Converging lens

Rays leave the tip of the object in all directions. All rays that enter a converging lens will be focussed at the tip of the image. Three rays in particular can be used to find that focus:

- The ray that enters the lens at its centre goes straight through without any change of direction.
- The ray that approaches the lens parallel to the axis passes through the **principal focus** when it leaves the lens.
- The ray that passes through the principal focus before entering the lens is parallel to the axis when it leaves the lens.

Any two of these could be used to find the tip of the image. The third acts as a double-check on the accuracy of your drawing.

# Diverging lens

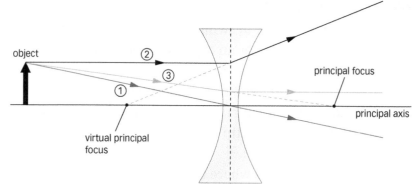

▲ Three special rays for working out images from diverging lenses.

These three rays can be used to find the position of the tip of the image formed by a diverging lens.

- The ray that enters the lens at its centre goes straight through without any change of direction.
- The ray that approaches the lens parallel to the axis appears to come from the virtual principal focus when it leaves the lens.
- The ray that is heading towards the principal focus when it meets the lens travels parallel to the axis when it leaves the lens.

The result is three rays coming out of the lens, spreading apart from each other. When these rays enter your eye, your brain works out where the rays appear to come from and places the tip of the image there. This is always somewhere behind the lens, where a screen would block out the light, so it is always a **virtual** image.

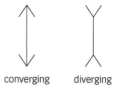

converging    diverging

▲ These symbols will represent lenses in the exam.

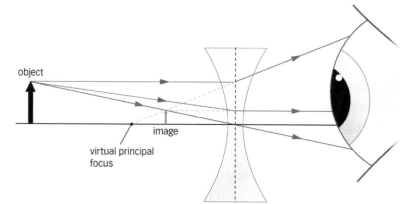

▲ The image is where the rays appear to come from.

## Questions

1   What happens to light that enters the centre of a lens?

2   Draw diagrams to show the effect of converging and diverging lenses on light rays parallel to the axis.

3   **H** A converging lens has a focal length of 5 cm. Describe the image formed by the lens of a 2-cm-high object placed 10 cm away from the lens.

## Magnification

The image formed by a lens and the object itself usually have different heights. You calculate the **magnification** produced by the lens with this equation:

$$\text{magnification} = \frac{\text{height of image}}{\text{height of object}}$$

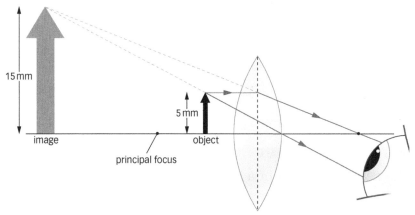

▲ Using a convex lens to give a magnification of 3.

Suppose you look at a 5 mm object with a magnifying glass. If the image is 15 mm high, what is the magnification?

$$\text{magnification} = \frac{\text{height of image}}{\text{height of object}} = \frac{15\,\text{mm}}{5\,\text{mm}} = 3$$

The magnification is often smaller than 1. This means that the image is diminished. For example, suppose that a convex lens produces an image that has a magnification of 0.7. What is the size of the image if the object is 5 cm high?

$$\text{magnification} = \frac{\text{height of image}}{\text{height of object}}$$

so magnification × height of object = height of image

$$\text{height of image} = 0.7 \times 5\,\text{cm} = 3.5\,\text{cm}$$

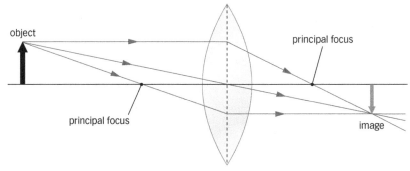

▲ This lens is providing a magnification of only 0.7.

### Revision objectives

- ✓ calculate the magnification of an object
- ✓ calculate the power of a lens
- ✓ describe the factors affecting the power of a lens
- ✓ use refraction to explain the factors affecting the power of a lens

### Student book references

**3.10** Magnification

**3.14** Lens power

### Specification key

- ✔ P3.1.3 e, f and i
- ✔ P3.1.4 e – g

# Power

The **power** of a lens tells you how good it is at converging or diverging light. You calculate it from this equation:

$$P = \frac{1}{f}$$

$P$ is the power of the lens in **dioptres** (D)
$f$ is the **focal length** of the lens in metres (m)

So if a converging lens has a focal length of 20 cm, what is its power?

$$P = \frac{1}{f} = \frac{1}{0.20\,m} = +5\,D$$

The power of a converging lens is always positive. Diverging lenses have negative powers.

## Increasing power

To increase the power of a lens you have to reduce its focal length. There are two ways of doing this:

• increase the **refractive index**
• increase the curvature (how fat the lens is).

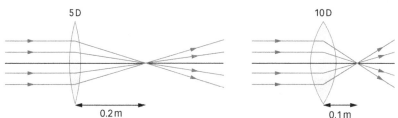

5 D                                    10 D

0.2 m                                    0.1 m

▲ Both of these lenses have the same refractive index, but different curvatures.

**H** A high-power lens changes the direction of light more than a low-power one. The refractive index of the material used affects how much the light changes direction on its way through the lens. So increasing the refractive index of the material is a useful alternative to increasing the curvature of the lens when you need to increase its power.

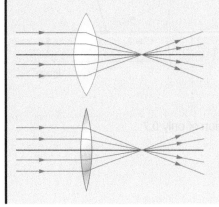

◀ Both lenses have the same power but different values of refractive index.

## Exam tip                AQA

Always remember to convert the focal length to metres before calculating power.

## Questions

1   State the factors that affect the power of a lens.

2   A converging lens has a focal length of 0.05 m. Calculate its power.

3   **H** A lens has a power of −10 D. State the type of lens and calculate its focal length.

## Structure

The human eye contains several important structures:

- Light enters through the transparent **cornea**.
- The **pupil** controls how much light reaches the lens.
- The **lens** focuses the light onto the retina.
- Cells in the **retina** detect the intensity and colour of the light.
- The **optic nerve** carries electrical signals from the retina to the brain.

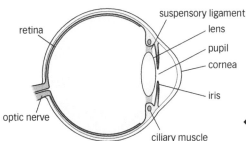

◀ A cross-section through the human eye.

## Brightness control

The cells on the surface of the retina can be damaged if they absorb too much light. The pupil automatically gets smaller in bright light to prevent this happening. The size of the pupil is altered by muscles in the **iris**. This is the coloured part of the eye. The pupil appears black because light is absorbed by the retina, not reflected.

## Focus control

Light is refracted by both the cornea and the lens. The lens is held in place by **suspensory ligaments**. The lens is soft, so the **ciliary muscles** can be used to make it fatter or thinner. This happens automatically to keep the image correctly focussed on the retina. This allows the eye to focus on objects from 25 cm (the **near point**) to infinity (the **far point**).

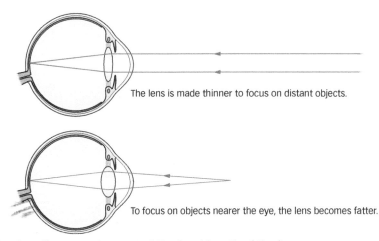

The lens is made thinner to focus on distant objects.

To focus on objects nearer the eye, the lens becomes fatter.

▲ The ciliary muscles control the focal length of the lens.

## Key words

retina, cornea, pupil, lens, optic nerve, iris, suspensory ligaments, ciliary muscles, near point, far point, short-sight, concave, long-sight, convex

## Exam tip    AQA

Remember that the eye contains *two* lenses: a fixed cornea and an adjustable lens.

## Questions

1  Draw a labelled diagram of the human eye.

2  Describe and explain the changes in the pupil when a bright light is shone into the eye.

3  **H** Describe and explain how short-sight can be corrected.

## Short-sight

Many people have **short-sight**. Their far point is much less than infinity – perhaps less than a metre. The image created by the lens is focussed in front of the retina. This means that distant objects appear blurred. It can be caused:

- by an eyeball that is too large
- when the lens cannot be made thin enough.

The condition can be corrected by placing a diverging (**concave**) lens in front of the eye – either as spectacles or as contact lenses.

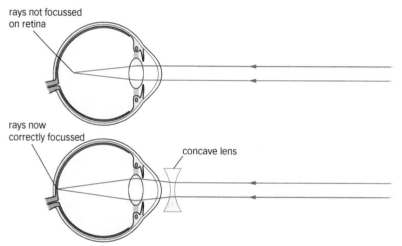

▲ A concave lens corrects short-sight.

## Long-sight

People who have **long-sight** have a near point that is much larger than normal. It can be caused:

- by an eyeball that is too small
- when a lens cannot be made fat enough.

The condition can be corrected by placing a converging (**convex**) lens in front of the eye – either as spectacles or as contact lenses.

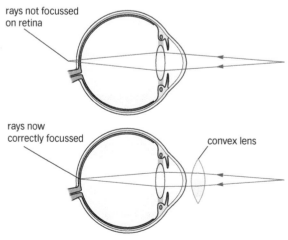

▲ A convex lens corrects long-sight.

## Common features

Many features of a camera match similar features in your eyes:

- a **convex lens** to catch light from an object and make an image
- an adjustable **iris** to control the amount of light in the image
- a **charge-coupled device (CCD)** or **film** to detect light in the image.

### Revision objectives

- ✔ describe how a camera forms an image
- ✔ compare the structures of a camera and an eye

### Student book references

**3.13** The camera

### Specification key

✔ P3.1.4 d

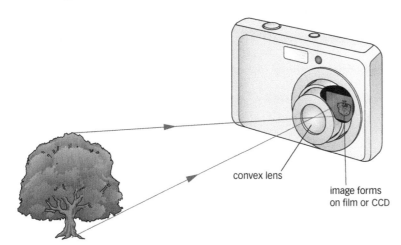

convex lens

image forms
on film or CCD

▲ The convex lens of the camera produces a real image of the object on the CCD.

## Camera focus

Both eyes and cameras have to produce focussed images of objects at a range of distances. The eye does this by altering the shape of the lens, making it fatter or thinner. The lens in a camera is hard, so has a constant shape. It therefore has to be moved towards or away from the CCD to bring images into focus on the CCD. As the object gets nearer to the camera, the lens has to be moved away from the CCD to keep the image in focus.

### Key words

convex lens, iris, charge-coupled device (CCD), film

### Questions

1 What are the functions of the lens and the CCD in a camera?

2 Draw a ray diagram to show how a camera forms an image.

3 **H** Describe and explain the differences in how an eye and a camera adapt to an object moving towards them.

### Exam tip

Make sure that you can explain the differences between a camera and an eye.

## Working to Grade E

1  Draw a diagram to show the refraction of light passing from glass into air.

2  Draw diagrams to show the effect of convex and concave lenses on parallel rays of light.

3  What is the difference between an object and an image?

4  A data projector produces an image that is **magnified** and **inverted**. What does this mean?

5  Some images are real. Others are virtual. What is the difference?

6  A magnifying glass has a magnification of 6. What does this mean?

7  Describe how the focal length of the human eye can be adjusted.

## Working to Grade C

8  A ray of light has angles of incidence and refraction of 35° and 25°, respectively, when it goes into plastic. Calculate the refractive index of the plastic.

9  Draw a ray diagram to show how a convex lens can form an inverted and diminished image of an object.

10  Draw a ray diagram to show how a convex lens can form a magnified, upright, and virtual image of an object.

11  Explain the adjustment required of a camera lens when a close object moves away from the camera.

12  A magnifying glass produces an image with a magnification of 5. If the object is 2 mm high, how high is the image?

13  Explain the function of the iris of a human eye.

14  With the help of a ray diagram, explain the type of lens required to correct long-sight.

15  Calculate the power of a convex lens with a focal length of 15 cm.

16  Add labels to this diagram of the human eye.

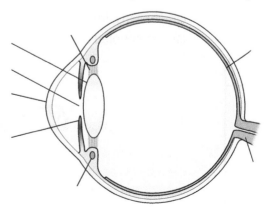

## Working to Grade A*

17  A convex lens has a power of +10 D. It is used to form an image of an object that is 5 cm from the lens. Use a ray diagram to calculate the magnification of the image.

18  Explain the different ways that eyes and cameras produce focussed images of objects at various distances away.

1   The ray diagram shows a lens being used to project an image of an object.

The diagram has been drawn to scale.

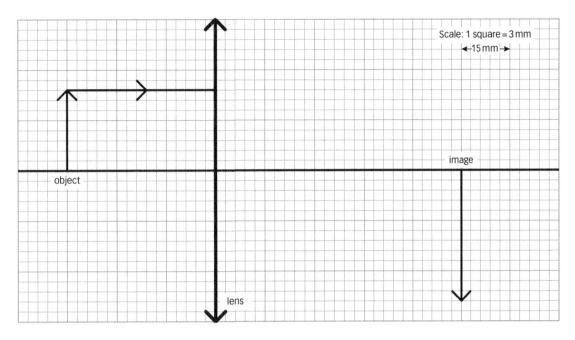

Scale: 1 square = 3 mm
←15 mm→

a   Use **three** words from the box to complete the sentence that describes the image.

| diminished | inverted | magnified | real | upright | virtual |
|---|---|---|---|---|---|

The image is ........................., ......................... and ......................... .

*(3 marks)*

b   i   The ray diagram shows one ray leaving the top of the object. Draw the ray after it has left the lens.

*(2 marks)*

ii   Use the ray diagram to calculate the power of the lens.

Write down the equation you use and then show clearly how you work out your answer.

.................................................................................................................................................

.................................................................................................................................................

.................................................................................................................................................

.................................................................................................................................................

.................................................................................................................................................

.................................................................................................................................................

power = ....................... D

*(3 marks)*

c   Draw **two** more rays from the top of the object that can be used to find the position of the image.

*(2 marks)*

***(Total marks: 10)***

**2** **a** The diagram shows the cross-section of a camera.

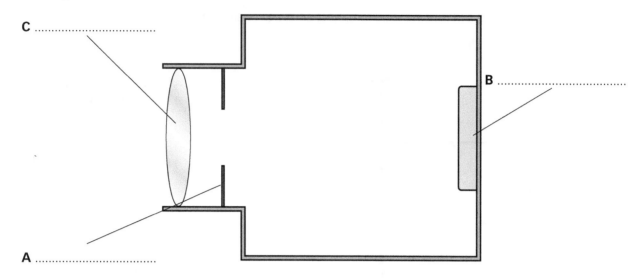

Use words from the box to label the parts **A**, **B**, and **C**.

| CCD | iris | lens | pupil | retina |
|-----|------|------|-------|--------|

(3 marks)

**b** Cameras and human eyes have many things in common.

Which one of these things is **not** common to both eyes and cameras?

Tick (✓) **one** box.

The shape of the lens can be altered. ☐

The image is focussed on a light detector. ☐

The amount of light in the image can be altered. ☐

(1 mark)

**c** The ray diagram shows the lens of a human eye. The principal focus is marked F.

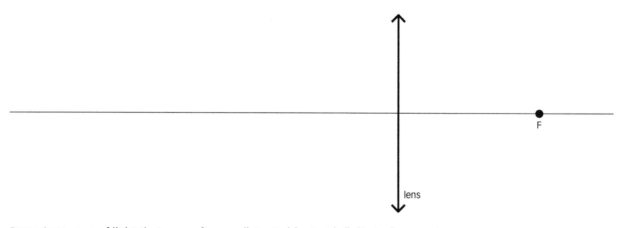

Draw **two** rays of light that come from a distant object at infinity to form an image.

(3 marks)

**(Total marks: 7)**

**3** The diagram shows a ray of light passing from air into glass.

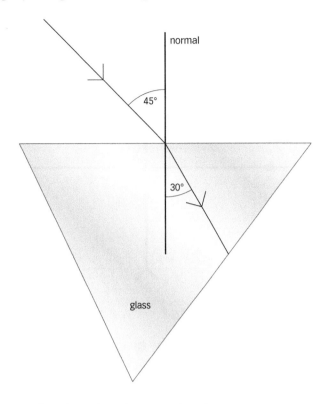

**a** Explain why the light changes direction where it enters the glass.

..............................................................................................................................................................................

..............................................................................................................................................................................

..............................................................................................................................................................................

..............................................................................................................................................................................

*(3 marks)*

**b** Calculate the refractive index of the glass.

Write down the equation you use and then show clearly how you work out your answer.

..............................................................................................................................................................................

..............................................................................................................................................................................

..............................................................................................................................................................................

..............................................................................................................................................................................

..............................................................................................................................................................................

refractive index = .....................

*(2 marks)*

**c** Continue the ray to show what it does when it passes out of the glass and back into the air.

You do not need to do any calculations.

*(2 marks)*

***(Total marks: 7)***

**4** The ray diagram shows the position of the virtual principal focus P of a concave lens.

The diagram is drawn to scale.

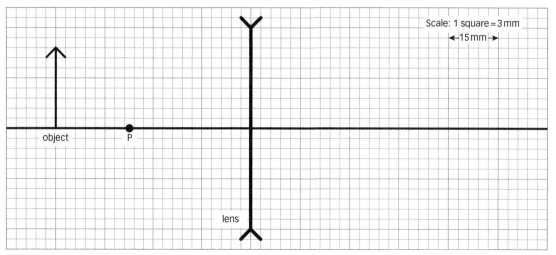

Scale: 1 square = 3 mm
←15 mm→

object    P

lens

**a** Calculate the power of the lens.

Write down the equation you use and then show clearly how you work out your answer.

...................................................................................................................................................................

...................................................................................................................................................................

...................................................................................................................................................................

...................................................................................................................................................................

...................................................................................................................................................................

lens power = ........................D

*(3 marks)*

**b** Draw **two** rays from the tip of the object to show how the image is formed.

Label the tip of the image with the word **image**.

*(3 marks)*

**c** Calculate the magnification of the image.

Write down the equation you use and then show clearly how you work out your answer.

...................................................................................................................................................................

...................................................................................................................................................................

...................................................................................................................................................................

...................................................................................................................................................................

...................................................................................................................................................................

magnification = ........................

*(3 marks)*

**(Total marks: 9)**

## Internal reflection

When light passes from one medium to another there is always some internal reflection as well as refraction. Some of the light reflects off the boundary, the rest changes direction as it passes through. If the light is travelling out of a denser medium at a large enough angle of incidence, then the light will only be reflected. All of it will reflect off the boundary, obeying the laws of reflection, with no loss of intensity. This **total internal reflection** (or **TIR**) only happens when the angle of incidence is greater than the **critical angle**.

### Revision objectives

- ✔ describe the conditions needed for total internal reflection
- ✔ describe how light passes along optical fibres
- ✔ give examples of the use of total internal reflection
- ✔ describe medical uses of lasers
- ✔ calculate the critical angle

### Student book references

**3.15** Total internal reflection

**3.16** Optical fibres and lasers

### Specification key

- ✔ P3.1.5

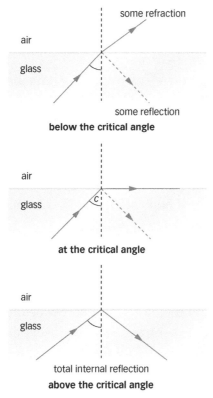

some refraction

air

glass

some reflection

**below the critical angle**

air

glass

*c*

**at the critical angle**

air

glass

total internal reflection

**above the critical angle**

▲ Increasing the angle of incidence can lead to total internal reflection.

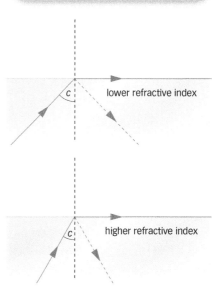

*c*   lower refractive index

*c*   higher refractive index

▲ The size of the critical angle depends on the refractive index.

---

### H  Critical angle

When the angle of incidence equals the critical angle $c$, the angle of refraction is 90°. This means that you can calculate it with this equation:

$$\text{refractive index} = \frac{1}{\sin c}$$

So what is the critical angle for glass of refractive index 1.6?

$$\text{refractive index} = \frac{1}{\sin c}$$

so $\sin c = \dfrac{1}{\text{refractive index}} = \dfrac{1}{1.6} = 0.625$

so $c = \sin^{-1} 0.625 = 39°$

## Optical fibres

An **optical fibre** is a narrow thread of very transparent glass or plastic. Light that enters one end at a shallow enough angle is unable to refract out into the air until it reaches the other end. The light totally internally reflects off the sides with no loss of energy. This happens even if the fibre is bent – the light has to follow the fibre.

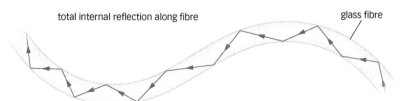

total internal reflection along fibre                    glass fibre

▲ Total internal reflection of light in an optical fibre.

Optical fibres have two main uses:
- They carry pulses of infrared over long distances, rapidly transferring large amounts of information in code for the internet and telephone communications.
- They allow photos to be taken deep inside the body with an **endoscope**. A bundle of fibres carries light into the body, usually through the mouth or anus. A lens collects reflected light and sends it back along another bundle of fibres, allowing the image to be viewed.

## Lasers

Optical fibres often use light from a **laser** as the source of their pulses. Laser light forms a very narrow beam that does not spread very much as it travels. Some lasers can produce very intense light that is useful for burning and cutting materials. Medical uses of lasers include:
- precision cutting out of unwanted growths, such as tumours and scar tissue
- sealing off blood vessels by cauterising them
- correcting eyesight by changing the shape of the cornea.

## Questions

1   Describe **two** applications of TIR.

2   What are the **two** conditions necessary for TIR to occur?

3    Calculate the critical angle for a glass of refractive index 1.4.

# Questions
## Total internal reflection, optical fibres, and lasers

### Working to Grade E

1   State **two** applications of lasers.

2   Describe an optical fibre and give **one** use for it.

### Working to Grade C

3   Complete this diagram to show TIR in an optical fibre.

glass

4   Why is it better to send pulses of laser light down an optical fibre instead of through the air for Internet communications?

5   A clear piece of plastic has a critical angle of 42°. Calculate its refractive index.

6   Explain the conditions required for TIR to take place.

7   Explain **one** use of lasers in surgery.

### Working to Grade A*

8   A clear piece of plastic has a refractive index of 1.3. Calculate the critical angle for the plastic.

**1** The diagram shows how a glass prism can be used to see over the top of a wall.

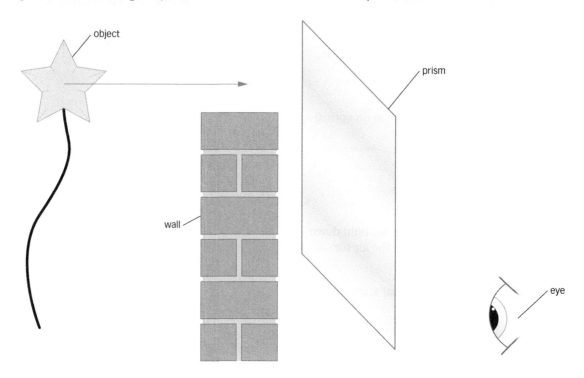

**a** The glass has a refractive index of 1.6. Calculate the critical angle for the glass.

Write down the equation you use and then show clearly how you work out your answer.

......................................................................................................................................................

......................................................................................................................................................

......................................................................................................................................................

......................................................................................................................................................

critical angle = ............... degrees

*(2 marks)*

**b** The diagram shows a light ray leaving the object. Continue the ray on the diagram to show how it reaches the eye.

*(3 marks)*

**c** Draw a ring around the correct answer to complete the following sentence.

The image seen by the eye is
| inverted | false |
|----------|-------|
| rotated | real |
| upright | virtual |
and

*(2 marks)*

***(Total marks: 7)***

## Designing an investigation and making measurements

In this module there are some opportunities to design investigations and make measurements. These include measurements of sizes and positions of images produced by lenses, through refraction of light.

Although you may be asked to demonstrate your investigative skills practically, you are also likely to be asked to comment on investigations done by others. The example below offers guidance in this skill area. It also gives you the chance to practise using your skills to answer the sorts of questions that may come up in exams.

## Investigating magnification of images

### Skill – Understanding the experiment

Nina tested the hypothesis that the magnification of an image produced by a convex lens is proportional to its distance from the centre of the lens.

She drew a line of length 25 mm on a piece of tracing paper and placed it in front of a lamp. She placed a white card screen some distance away and moved a convex lens until it produced a focussed image of the line on the screen. Nina used a ruler to measure the height of the image and the distance of the screen from the centre of the lens. She repeated the experiment four times for different placements of the screen.

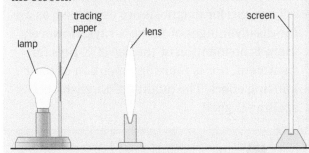

Nina's results are in the table.

| Height of object in mm | Distance from paper to screen in cm | Height of image in mm | Distance from lens to screen in cm |
|---|---|---|---|
| 25 | 120 | 9 | 34 |
| 25 | 135 | 98 | 109 |
| 25 | 103 | 54 | 69 |
| 25 | 92 | 31 | 49 |

1  Identify the independent and dependent variables.

In an investigation:
- The independent variables are the ones that are changed by the scientist.
- The dependent variable is the one that is measured for each change of the independent variables.

2  Suggest important control variables for the investigation.

A control variable is one that the investigator thinks might affect the outcome, so they try to keep this variable the same all the way through.

### Skill – Using data to draw conclusions

3  How can Nina use the data to test her hypothesis?

Nina's hypothesis concerns how the image height depends on its distance from the lens. So she should plot a scatter graph of these two quantities and look for a correlation between them. Alternatively she could divide one variable by the other and look to see if the result is the same every time.

4  What should Nina do if her results don't support her hypothesis?

If the data does not support the hypothesis, then:
- the data needs to be checked by repeating the experiment, or
- the control variables need to be thought about more carefully – perhaps one has been overlooked, or
- the model used to make the hypothesis needs to be examined more carefully – Nina probably used ray diagrams to make her prediction.

### Skill – Evaluating the experiment

5  Give reasons why Nina might get different results when she repeats the experiment.

There are several reasons why results of an experiment are different when it is repeated.
- Human error may change values in a random way, giving a spread of results.
- The limited resolution of the measuring instruments makes it impossible to set up the apparatus exactly the same, every time.
- A zero-error in the measuring instrument may give rise to a systematic error.

# AQA Upgrade

## Answering an extended writing question

QUESTION

*In this question you will be assessed on using good English, organising information clearly, and using specialist terms where appropriate.*

**1** Explain the advantages and disadvantages of using X-rays for the diagnosis and treatment of cancer tumours.

*(6 marks)*

---

**G–E**

Well, X-rays aren't much good for finding tumours as they aren't made of bone, only bone shows up in an X-ray, so you wouldn't see the tumour, it would look just like the rest of the skin and mussel. But X-rays are good at killing stuff, so once you kno where the tumour is you can focuss the X-rays onto it and fry it so that it is all killed and not dangerous any more.

**Examiner:** This answer is typical of a grade-G candidate. It is worth just 1 mark.

The candidate has mentioned only two correct facts about X-rays and has completely ignored the use of CT scans. There is no attempt to explain either fact. X-rays cannot be focussed. The sentences are rather long, with poor spelling and repetition.

---

**D–C**

X-rays stops doctors from having to open the patient up, as they can be used to make images of slices through the body. A CT scan sends a thin beam of X-rays through the body to a CCD detector on the other side. A computer puts together the information from sending the X-rays at lots of different angles, making an image on the screen. So doctors can see the tumour even if it is deep inside. However, X-rays are dangerous and can give you cancer if you don't already have it, so this is only worth doing if you know that you already have it.

**Examiner:** This answer is typical of a grade-D candidate. It is worth just 3 marks.

There is a good description of the CT scan process, but no reference to the variable absorption of X-rays by different tissues. The advantages for diagnosis are explained, as are the disadvantages of using X-rays. However, there is no mention of the use of X-rays for treatment, nor is there any mention of their ionising effect. The quality of English and spelling is good.

---

**B–A\***

X-rays are high energy electromagnetic waves with enough energy to ionise atoms in their path. This means that they can pass deep into the body before being absorbed. High density materials absorb better than low density ones, so the intensity of an X-ray beam which passes through part of the body depends on the organs in its path – including a tumour. So if the X-rays are fired through the same part of the body at lots of different angles a computer can create an image of that part. This allows doctors to detect tumours without having to cut open the body (which carries an infection risk). Beams of X-rays from lots of different directions can then be used to kill the cells of the tumour, provided it is where the beams cross. One problem is that other cells may have their DNA damaged by the ionisation, making them into cancer.

**Examiner:** This answer is typical of a grade-A candidate. It is worth 5 marks.

The information is well ordered, starting with the properties of X-rays, which can be used to explain the techniques described in the rest of the response. Both diagnosis and treatment have been discussed. There is no mention of how X-rays are detected, and the risk–benefit analysis of treatment is a little shallow. The use of English is excellent.

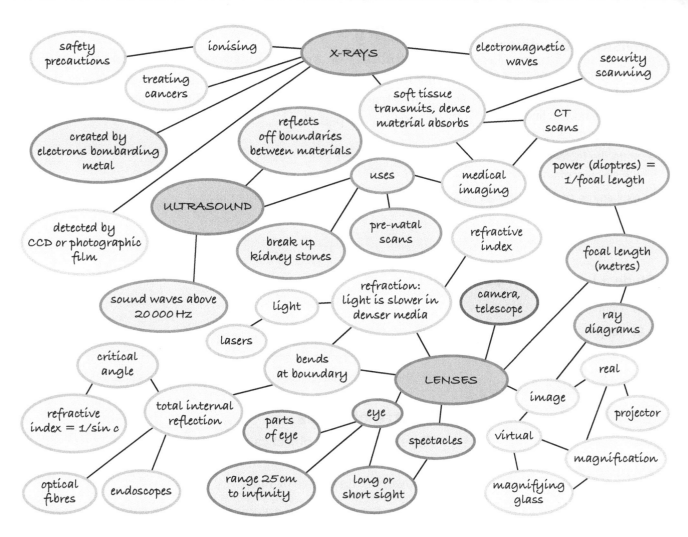

## Revision checklist

- X-rays are electromagnetic waves with a short wavelength. They are ionising, so potentially dangerous.
- Ultrasound waves are sound waves with a frequency above 20 000 Hz. Ultrasound reflects off boundaries between materials.
- Refraction is the bending of light when it moves from one medium to another. Lenses use refraction to form an image.
- Refractive index is a measure of the speed of light through a material compared with the speed of light in a vacuum.
- The nature of an image is determined by its distance from the lens, its size, its orientation, and whether it is real or virtual.
- The image produced by converging lenses is real and inverted, or virtual, upright, and magnified.
- The image produced by diverging lenses is virtual, upright, and diminished.
- Magnification = image height/object height.
- The human eye has a range of vision from 25 cm to infinity. Features of the eye: retina; cornea; iris; pupil; ciliary muscles; suspensory ligaments.
- Short-sight is inability to focus on far away objects, corrected by diverging lenses.
- Long-sight is inability to focus on near objects, corrected by converging lenses.
- In a camera a converging lens focuses light onto photographic film or a charge-coupled device (CCD) to produce a real image.
- More powerful lenses have shorter focal lengths. Lens power is measured in dioptres. Power = 1/focal length.
- If the angle of light travelling from a denser medium towards a less dense medium exceeds the critical angle, total internal reflection (TIR) occurs.
- Optical fibres channel light through total internal reflection, and are used for communication and in endoscopes.
- Laser beams remain very narrow even over long distances. Their intensity means they can be used to cut, etch, or cauterise objects (including in laser eye surgery).

## Revision objectives

✓ understand the meaning of the term 'centre of mass'
✓ describe how to find the centre of mass of an object
✓ state the factors that affect the stability of an object
✓ explain the conditions for an object to topple

# Stability

The **stability** of an object tells you how far it can be tilted before it falls over. So a pencil standing on its end is less stable than a traffic cone. This is because the cone has a much wider base than the pencil.

▲ The pencil is much easier to topple than the traffic cone.

# Centre of mass

The stability of an object can be increased by:
• increasing the width of its base
• lowering its centre of mass.

The **centre of mass** of an object is the point at which all of its **weight** appears to act. You can find this point by suspending the object from two different points: the object always settles with the centre of mass directly below the suspension point.

▲ His centre of mass is directly below his right hand.

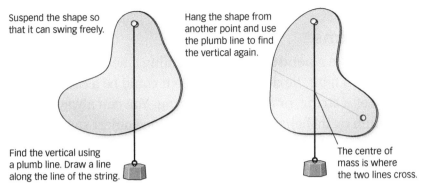

Suspend the shape so that it can swing freely.

Hang the shape from another point and use the plumb line to find the vertical again.

Find the vertical using a plumb line. Draw a line along the line of the string.

The centre of mass is where the two lines cross.

▲ Finding the centre of mass of an object.

## Symmetry

When the mass is evenly spread over a symmetrical object, then the centre of mass is somewhere on the axis of **symmetry**.

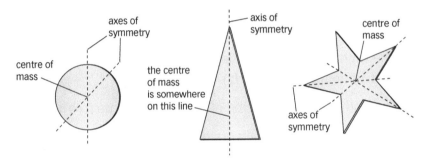

axes of symmetry

centre of mass

axis of symmetry

the centre of mass is somewhere on this line

centre of mass

axes of symmetry

▲ Using symmetry to find centres of mass.

## H Explaining stability

An object always settles with the **line of action** of its weight acting through the base. A slight tilt leaves the centre of mass above the base, so there is a **moment** returning it to its previous position. A large tilt puts the centre of mass outside the base, giving a moment that increases the tilt even more – so the object falls over.

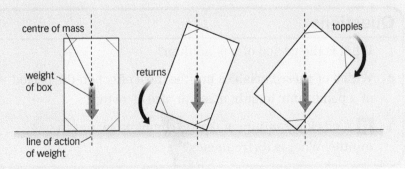

centre of mass

weight of box

returns

topples

line of action of weight

## Questions

1   Where is the centre of mass of a solid cube?

2   State **two** ways of increasing the stability of an object.

3   H Explain why a pencil always topples over when it is placed on a table point down.

## Revision objectives

- ✓ describe the motion of a pendulum
- ✓ explain how the time period of a pendulum is related to its length

## Student book references

**3.18** Pendulums

## Specification key

✓ P3.2.1 d – e

## Key words

pendulum, period, frequency

# Pendulums

You can make a **pendulum** by suspending a heavy object on the end of a light, flexible support. So it could be a conker on the end of a string, or a child on a swing. You can always spot a pendulum by its behaviour – the suspended object swings backwards and forwards when you give it a brief push.

# Period

Pendulums have been used in clocks for hundreds of years. This is because the time that the object takes to make one complete swing (the **period**) mostly depends on the length of its suspension. Changing the mass of the object or the size of its swing makes hardly any difference to its period. So the **frequency** of the pendulum remains steady – the number of oscillations per second does not change.

$$T = \frac{1}{f}$$

$T$ is the time period in seconds (s)

$f$ is the frequency in hertz (Hz)

So if a pendulum has a period of 0.25 s, what is its frequency?

$$T = \frac{1}{f} \text{ so } f = \frac{1}{T} = \frac{1}{0.25\,\text{s}} = 4.0\,\text{Hz}$$

## Exam tip AQA

Remember that the period is the time for the pendulum to swing there and back again.

## Questions

1  What is the period of a pendulum?

2  Which of these variables has the most effect on the period of a pendulum: length, mass, or size of swing?

3  H A clock pendulum makes 15 complete swings in one minute. What is its frequency?

## Turning forces

Forces are often used to make things turn. The effect of a turning force depends on three things:

• the size of the force
• the direction of the force
• how far it acts from the pivot.

The **pivot** is the point on the object that does not move when it rotates.

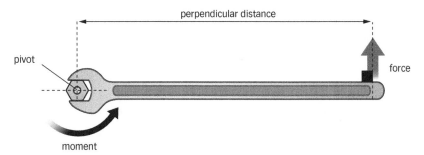

▲ A force on the spanner handle provides a moment about the nut. These factors are put together in the idea of a moment.

## Calculating moments

You calculate the size of a **moment** with this equation:
$$M = F \times d$$

$M$ is the moment of the force in netwon-metres (Nm)
$F$ is the force in newtons (N)
$d$ is the perpendicular distance in metres (m)

The **perpendicular distance** takes account of the direction of the force. It is the distance from the **line of action** of the force from the pivot. So what is the moment applied to this spanner?
$$M = F \times d = 60\,\text{N} \times 0.2\,\text{m} = 12\,\text{Nm}$$

Moments have direction as well as a size. The moment on the spanner is **anticlockwise**. Reversing the direction of the force would give a **clockwise** moment.

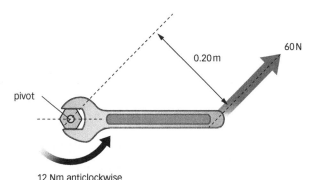

12 Nm anticlockwise

▲ The perpendicular distance is how far the force's line of action is from the pivot.

### Revision objectives

✓ understand that the turning effect of a force is a moment

✓ calculate the size of a moment

✓ explain the principle of moments

✓ calculate forces and distance for objects balanced on a pivot

### Student book references

**3.19** Moments

**3.20** Principles of moments

### Specification key

✓ P3.2.2 a – d

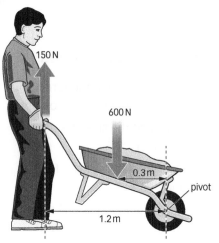

The wheelbarrow is balanced.

# Principle of moments

The **principle of moments** says that an object balances when the moments acting on it cancel each other out.

The wheelbarrow on the left is balanced. The clockwise moment from the man is 150 N × 1.2 m = 180 Nm. The anticlockwise moment from the weight of the wheelbarrow is 600 N × 0.3 m = 180 Nm. They cancel out.

## H Calculating with moments

Sam and Jo sit on a seesaw. The seesaw balances when they sit as shown. If Sam has a weight of 600 N, what is Jo's weight?

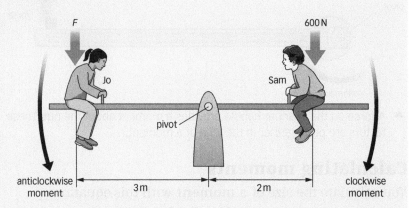

$$\text{anticlockwise moment} = \text{clockwise moment}$$
$$F \times 3\,\text{m} = 600\,\text{N} \times 2\,\text{m} = 1200\,\text{Nm}$$
$$F \times 3\,\text{m} = 1200\,\text{N} \text{ so } F = \frac{1200\,\text{Nm}}{3\,\text{m}} = 400\,\text{N}$$

## Key words

pivot, moment, perpendicular distance, line of action, anticlockwise, clockwise, principle of moments

## Questions

1  What is the direction of the moment applied to the door?

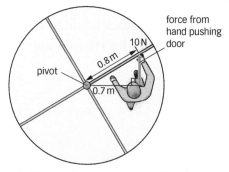

2  How big is the moment applied to the door?

▲ Turning force on a revolving door.

3  **H** Bill and Ted sit on a balanced seesaw. Bill has a weight of 600 N and sits 2.4 m to the left of the pivot. Ted has a weight of 800 N. Where does he sit?

## Force multiplier

This bottle opener uses a **lever** as a **force multiplier**.

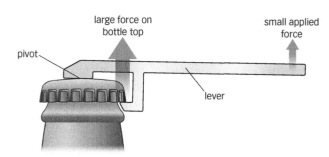

large force on
bottle top

small applied
force

pivot

lever

▲ A bottle opener in action.

A small force is applied to the lever a long way from the **pivot**. A much larger force is exerted by the lever on the bottle top. The two **moments** have to be the same, so the force closer to the pivot must be larger.

Here are some other devices that increase the size of your force:

- a crowbar for removing nails from wood
- scissors for cutting string and paper
- a screwdriver to open a tin of paint.

They all work in the same way. The lever is pivoted near one end and your force is applied to the other end. This allows a much larger force at some point close to the pivot.

### Exam tip

AQA

Always start analysing a lever by identifying the pivot.

### Questions

1   Where is the pivot of this lever?

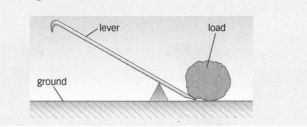

lever

load

ground

2   Label the force applied to the lever and the force from the lever on the load.

3   **H** Explain how increasing the length of the lever makes it easier to raise the load.

### Revision objectives

✔ state some examples of levers

✔ explain how a lever multiplies a force

### Student book references

**3.21** Levers

### Specification key

✔ **P3.2.2 e**

applied
force

pivot

greater force applied to remove nail

▲ A crowbar.

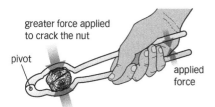

greater force applied
to crack the nut

pivot

applied
force

▲ A nutcracker.

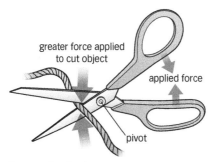

greater force applied
to cut object

applied force

pivot

▲ A pair of scissors.

### Key words

**lever, force multiplier, pivot, moment**

## Revision objectives

- recall that liquids are almost incompressible
- explain that pressure in liquids is transmitted equally in all directions
- use the equation linking pressure, force, and area
- explain how a hydraulic system multiplies a force

## Student book references

**3.23** Pressure in liquids

**3.24** Hydraulics

## Specification key

- P3.2.3

# Pressure

The weight of the liquid in this cylinder exerts a downwards **pressure** on its base. You can calculate pressure with this equation:

$$P = \frac{F}{A}$$

$P$ is the pressure on a surface in pascals (Pa)
$F$ is the force acting on the surface in newtons (N)
$A$ is the area of the surface in metres squared (m²)

So if the base has an area of 0.01 m² and holds up water of weight 50 N, what is its pressure?

$$P = \frac{F}{A} = \frac{50\,\text{N}}{0.01\,\text{m}^2} = 5000\,\text{Pa}$$

A liquid always exerts a pressure at right angles to the surface in contact with it. So the side walls at the base of the cylinder also have a pressure of 5000 Pa, but in a horizontal direction.

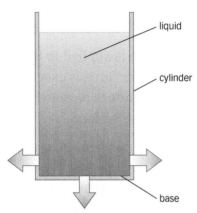

▲ The weight of the liquid pushes against the walls and base of the cylinder.

# Hydraulic systems

Any extra pressure applied to a liquid raises the pressure at any surface in contact with it. This is exploited by **hydraulic systems**, where a force applied at one point is transmitted through liquid in a pipe to a different point. This is widely used for car brakes:

- A force is applied to the brake pedal.
- The pressure of the oil in the cylinder increases.
- The pressure of the oil in the pipes also increases.
- The oil applies a force on the brake pads.

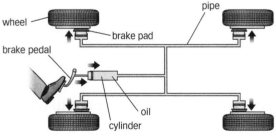

▲ Pressure from the brake pedal passes through the oil in the pipes to the brake pads.

## Incompressible

Like all liquids, the oil used in a hydraulic system is **incompressible**. Its particles are already very close to each other, so when the pressure increases, its volume hardly changes. So any change of pressure rapidly spreads through the liquid.

## Force multipliers

Hydraulic systems do more than allow a force applied at one point to be delivered to a different point some distance away. They also act as **force multipliers**, making that force bigger or smaller.

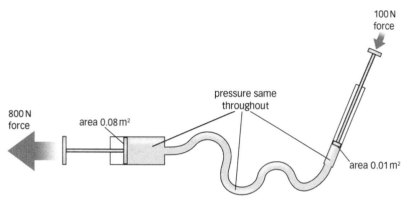

▲ A hydraulic system as a force multiplier.

This is how this hydraulic system works:
- A force of 100 N is applied to the oil over an area of 0.01 m² in the narrow cylinder.
- This raises the pressure of the oil by:
$$P = \frac{F}{A} = \frac{100\,\text{N}}{0.01\,\text{m}^2} = 10\,000\,\text{Pa}$$
- The pressure of the oil is the same in both cylinders.
- For the wider cylinder:
$$P = \frac{F}{A} = \text{so } F = P \times A = 10\,000\,\text{Pa} \times 0.08\,\text{m}^2 = 800\,\text{N}$$

So the force at the output of the system is eight times larger than the force applied at the input.

## Questions

1   A force of 500 N acts on an area of 0.25 m². What pressure does it exert?

2   Why is oil used in hydraulic systems instead of air?

3   **H** A force of 600 N is applied to a piston of area 0.02 m². If that pressure is transmitted through a liquid to another piston of area 0.08 m², what force is exerted on it?

### Exam tip

The pressure is the same everywhere in a hydraulic system, always pushing outwards.

## Working to Grade E

1   Where is the centre of mass of a box that is an empty cube?
2   A pendulum has a frequency of 0.5 Hz. How long does it take for 10 complete swings?
3   Calculate the moment of a force of 45 N applied at a distance of 0.8 m from a pivot.
4   Identify the clockwise and anticlockwise moments.

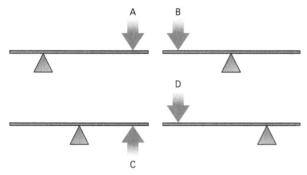

5   Give **two** examples of levers used at home.
6   Calculate the pressure when a force of 600 N is applied over an area of 0.3 m².

## Working to Grade C

7   Explain how you would find the centre of mass of a tennis racket.
8   State **two** factors that do **not** have much effect on the period of a pendulum.
9   Explain why this system is balanced.

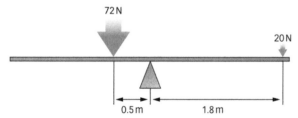

10  State **two** ways of increasing the stability of an object.
11  Why is there the same pressure on both pistons A and B?

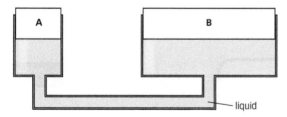

## Working to Grade A*

12  Calculate the force on piston B.

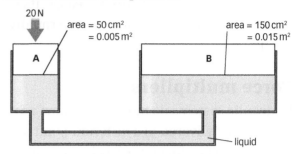

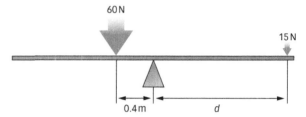

13  Calculate the value of *d* for the seesaw to balance.

14  Explain why nutcrackers allow you to crack a nut that you couldn't crack with your bare hands.

**1** The diagram shows Jack and Jill balancing on a seesaw.

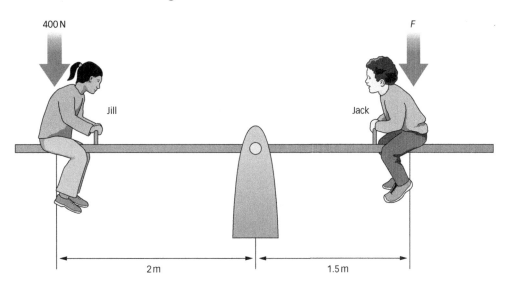

400 N

F

Jill

Jack

2 m

1.5 m

**a** Label the pivot of the seesaw.

*(1 mark)*

**b** Calculate the moment produced by Jill's weight.

Write down the equation you use and then show clearly how you work out your answer. Give the unit for your answer.

.............................................................................................................................................................

.............................................................................................................................................................

.............................................................................................................................................................

.............................................................................................................................................................

.............................................................................................................................................................

moment = .....................................

*(4 marks)*

**c** Draw a ring around the correct answer to complete the sentences.

Jill applies a moment in the | clockwise / downwards / anticlockwise | direction.

Jack's moment must be | the same as / smaller than / greater than | Jill's moment to balance the seesaw.

*(2 marks)*

**d** Calculate the force produced by Jack's weight.

Write down the equation you use and then show clearly how you work out your answer.

..........................................................................................................................................

..........................................................................................................................................

..........................................................................................................................................

..........................................................................................................................................

..........................................................................................................................................

force = ............................N

*(3 marks)*

**e** Jack owns a table tennis bat.

Describe how he could find the centre of mass of his table tennis bat.

..........................................................................................................................................

..........................................................................................................................................

..........................................................................................................................................

..........................................................................................................................................

..........................................................................................................................................

..........................................................................................................................................

..........................................................................................................................................

*(4 marks)*

***(Total marks: 14)***

## Changing direction

The string attached to the object in the diagram below makes it follow a circular path as it moves. This means that the direction of the object's motion has to keep on changing. So even if it keeps a constant **speed**, its **velocity** (which is fixed by both speed and direction) is continually changing. A changing velocity means that the object is accelerating.

$$\text{acceleration} = \frac{\text{change of velocity}}{\text{time taken for change}}$$

## Centripetal force

An object can only accelerate if there is a resultant force acting on it. The force that makes an object move in a circular path is called **centripetal force**, and it always has to act towards the centre of the circle. This is because the object is continually accelerating towards the centre of the circle. If there isn't a centripetal force on an object, it cannot move in a circle.

### Revision objectives

- ✔ understand that objects moving in a circle are accelerating
- ✔ describe centripetal force
- ✔ explain the effect of mass, speed, and path radius on centripetal force

### Student book references

**3.25** Circular motion

**3.26** Size of centripetal force

### Specification key

- ✔ P3.2.4

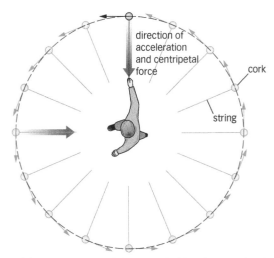

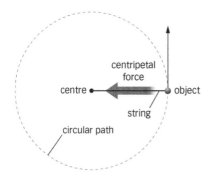

▲ As the object moves around the circle it keeps on changing direction.

▲ The centripetal force on the cork is provided by the tension in the string.

▲ Friction between the tyres and the road provides the centripetal force for this car as it goes around the bend.

# Changing centripetal force

The size of the centripetal force on an object can be increased by:

- increasing the **mass** of the object
- increasing the speed of the object
- decreasing the radius of the circular path.

The mass affects the centripetal force because force = mass × **acceleration**.

Making the object go faster means that its direction changes more in the same time, so it has a greater rate of change of velocity.

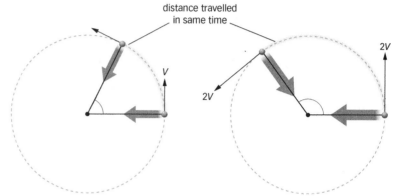

▲ Increasing the speed makes the direction change faster.

Making the circular path smaller also results in a greater change of direction in the same time. So there is a greater rate of change of velocity.

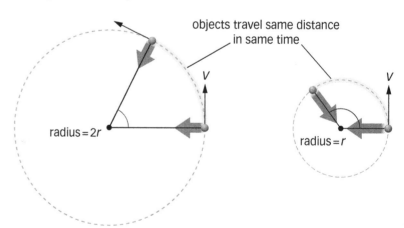

▲ Making the circle smaller makes the direction change faster.

## Exam tip

Remember that if an object moves in a circle the resultant force **must** point to the centre of the circle.

## Key words

speed, velocity, centripetal force, mass, acceleration

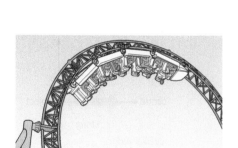

## Questions

1 What is the direction of the resultant force on the first rollercoaster car in the picture?

2 Explain why an object moving in a circle must have a centripetal force.

3 **H** You are sitting in a moving car when it swerves to the right. Explain why you appear to be pushed to the left-hand side of the car.

## Making magnetic fields

Whenever charge flows through a wire to make a **current**, a **magnetic field** appears around the wire. The shape of the field lines depends on the shape of the wire. The field lines are concentric circles for a straight wire, and like those of a bar magnet for a wire coiled into a **solenoid**.

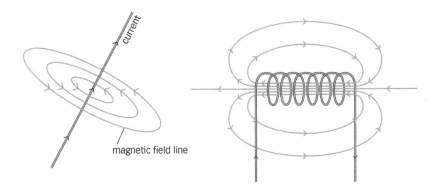

magnetic field line

## Motor effect

When magnets are placed close to each other, forces act between them. So when there is a current in a wire placed near to bar magnets, there is a **force** on the wire. This is called the **motor effect**. The size of the force can be changed by:

• increasing the size of the current
• increasing the strength of the bar magnets.

If the wire runs parallel to the magnetic field, it will not experience a force.

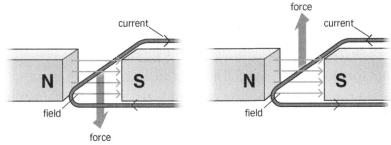

▲ There is a force on a current in a magnetic field.

The direction of the force is at right angles to both the current and the magnet's field, as shown by **Fleming's left-hand rule**. Changing the direction of either the current or the field changes the direction of the force – but changing the direction of both at the same time has no effect.

### Revision objectives

✓ understand that a current in a wire produces a magnetic field

✓ describe the force produced by the motor effect

✓ describe how to change the size and direction of this force

✓ describe and explain some applications of the motor effect

### Student book references

**3.27** Electromagnets and the motor effect

**3.28** Using the motor effect

### Specification key

✓ P3.3.1

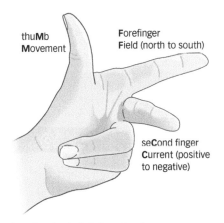

thuMb
Movement

Forefinger
Field (north to south)

seCond finger
Current (positive to negative)

▲ Fleming's left-hand rule.

# Electric motors

Electric motors are everywhere – from high-speed trains to vibrators in mobile phones. They use the motor effect to make a **coil** of wire spin round between a pair of magnets. Current from the power supply enters the coil via a pair of fixed **brushes** touching a **split-ring commutator**. The current in the coil interacts with the magnetic field to make a pair of equal forces which act in opposite directions. These forces are on different sides of the coil, so they make it turn. Each time the loop is vertical, the current is reversed. The side of the loop that was pushed up is now pushed down. The brushes and commutator act as a switch to keep the coil spinning in the same direction.

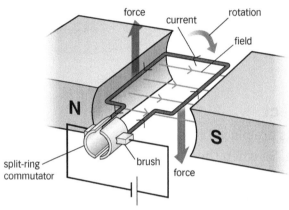

▲ A simple electric motor.

# Meters and speakers

An analogue ammeter measures current by passing it through a coil between magnets. The arrangement is very similar to that of an electric motor, with a spring to keep the coil in position. Increasing the current increases the force on the coil, twisting it round more against the spring.

**Loudspeakers** use the motor effect to make a paper cone vibrate rapidly. Alternating current is fed into the coil fastened to the back of the cone. The direction of the force from the magnet in the coil depends on the direction of the current. So as the current changes direction, so does the force.

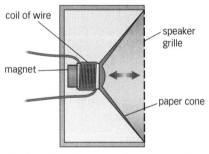

▲ Loudspeakers use the motor effect.

## Questions

1 Describe the magnetic field around a straight wire that has a current in it.

2 Explain why a solenoid can be used to make a magnet that can be turned on and off.

3  Explain **four** ways of making an electric motor spin round faster.

## Electromagnetic induction

Suppose a loop of wire sits in a **magnetic field**. Whenever the loop is moved so that part of it passes through the field lines, a **potential difference** is **induced** in the circuit.

The potential difference is only present during the movement of the loop, while it cuts the field lines. As soon as the movement stops, the potential difference disappears.

## Magnets and coils

Moving a magnet towards and away from a coil of wire induces a potential difference across the ends of the coil. The sign of the potential difference is reversed when the motion of the magnet is reversed. The potential difference disappears as soon as the magnet stops moving. This is because the field lines are no longer cutting through the wires of the coil.

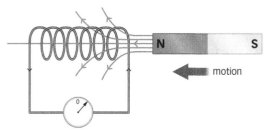

▲ Electromagnetic induction with a coil of wire and a magnet.

## Transformers

A **transformer** has two coils of wire wound on a single **iron core**. The **primary coil** is connected to a supply of **alternating current**, so that it becomes an electromagnet. Current in the primary coil creates magnetic field lines that pass through the core. As the current changes direction, so does the direction of the magnetic field. Each change of magnetic field direction in the core induces a potential difference across the **secondary coil**.

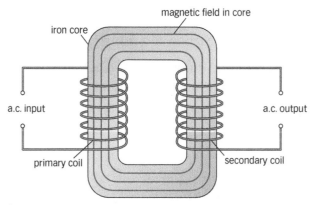

▲ A transformer.

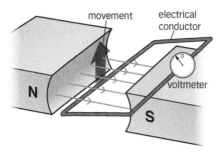

▲ The voltmeter gives a non-zero reading while the loop is moving.

## Key words

magnetic field, potential difference, induced, transformer, iron core, primary coil, alternating current, secondary coil

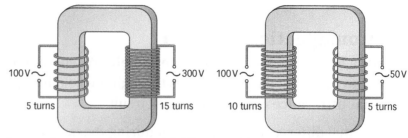

▲ Transformers make potential differences bigger or smaller.

## Changing potential difference

Transformers are widely used to change the potential difference of an alternating current. You can calculate the change with this equation:

$$\frac{V_p}{V_s} = \frac{n_p}{n_s}$$

$V_p$ is the potential difference across the primary coil in volts (V)
$V_s$ is the potential difference across the secondary coil in volts (V)
$n_p$ is the number of turns of wire on the primary coil
$n_s$ is the number of turns of wire on the secondary coil

Suppose a transformer is used to reduce the 230 V mains supply to 12 V. How many turns of wire should the secondary coil have if the primary has 4600 turns?

$$\frac{V_p}{V_s} = \frac{n_p}{n_s} \text{ so } n_s \times \frac{V_p}{V_s} = n_p \text{ and } n_s \times V_p = n_p \times V_s$$

$$\text{therefore } n_s = n_p \times \frac{V_p}{V_s} = 4600 \times \frac{12\,\text{V}}{230\,\text{V}} = 240$$

Notice that the ratio of the turns of wire on the two coils is the also the ratio of the potential differences across them.

## Exam tip AQA

Remember that a potential difference is **only** induced across the ends of a coil while the magnetic field inside it is changing.

## Questions

1 Describe the construction of a transformer.

2 There is a potential difference of +0.05 V across a coil while a magnet is being pushed into it. What is the potential difference while the magnet is being pulled out of the coil?

3 **H** A transformer has 9200 turns in its primary coil and 360 turns in its secondary coil. What is the potential difference across the primary when the secondary produces a.c. at 9 V?

# Electricity supply

Much of the electrical energy you use is generated as **alternating current** in power stations. This has to be raised to a high voltage by a **step-up** transformer before passing through transmission lines to where it is needed. A **step-down** transformer reduces the voltage of the alternating current to a safer value of 230 V before it arrives at your home. Both of these transformers can have a very high **efficiency**, so nearly all of the energy transferred into the **primary** coil is transferred out of the **secondary** coil. You can use this equation to calculate the rate of energy transfer:

$$P = V \times I$$

*P* is the rate of energy transfer for the coil in watts (W)
*V* is the potential difference across the coil in volts (V)
*I* is the alternating current in the coil in amperes (A)

A transformer that is 100% efficient will have the same **power** for both coils; all the energy transferred into the primary coil is transferred out of the secondary coil. You can therefore do calculations with this equation:

$$V_p \times I_p = V_s \times I_s$$

$V_p$ is the potential difference across the primary coil in volts (V)
$I_p$ is the alternating current in the primary coil in amperes (A)
$V_s$ is the potential difference across the secondary coil in volts (V)
$I_s$ is the alternating current in the secondary coil in amperes (A)

Suppose a step-down transformer reduces the potential difference from 11 kV to 430 V. If the current drawn from the secondary coil is 100 A, how much current enters the primary coil?

$$V_p \times I_p = V_s \times I_s \text{ so } I_p = \frac{V_s}{V_p} \times I_s = \frac{430\,\text{V}}{11\,000\,\text{V}} \times 100\,\text{A} = 3.9\,\text{A}$$

## Revision objectives

- ✔ use the transformer power equation
- ✔ compare transformers used for different applications
- ✔ know that switch mode transformers operate at high frequency
- ✔ describe the advantages of using a switch mode transformer

## Student book references

**3.31** Transformer power equation

**3.32** Switch mode transformers

## Specification key

✔ P3.3.2 e – k

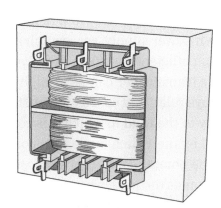

▲ Both coils transfer electrical energy at the same rate through the iron core. The low-voltage coil has to have thicker wire because it carries a larger current than the high-voltage coil.

## Switch mode transformer

Many devices plug straight into the mains supply. These include cookers, heaters, washing machines, and lights. They are designed to use alternating current at 230 V. Some devices, such as phones, computers, and games can only work on **direct current** at a low potential difference, perhaps 6 V. These devices are usually connected to the mains supply through a **switch mode transformer**.

▲ The switch mode transformer is built into the mains plug of this mobile phone charger.

The switch mode transformer does two things:
- It uses a small transformer to reduce the potential difference of the mains supply.
- It converts alternating current from the mains supply into direct current.

The transformer operates at a very high **frequency**. This allows it to be much smaller and lighter than a transformer that deals with 50 Hz alternating current.

## Efficiency

The electronic controls inside a switch mode transformer require their own power supply, so a switch mode transformer can never have an efficiency of 100%. However, the efficiency of a switch mode transformer can be as high as 95%. This means that it stays relatively cool as it transfers energy from the mains supply to the **load** to which it is connected. The current supplied by the switch mode transformer is determined by the potential difference across the load and its resistance. As soon as the load is removed from the switch mode transformer, this current drops to zero, so the only current drawn from the mains supply is the tiny amount for the control electronics.

### Key words

alternating current, step-up, step-down, efficiency, primary, secondary, power, direct current, switch mode transformer, frequency, load

### Exam tip

AQA

Show all the steps in your working out when doing transformer calculations. This allows you to spot mistakes.

### Questions

1 What does a switch mode transformer do?

2 What is the difference between a step-up and a step-down transformer?

3 **H** A step-up transformer has a primary coil of 500 turns and a secondary coil of 6 000 turns. If the current in the primary coil is 4 A, what is the current in the secondary coil?

166

# Questions
## Circular motion and power

1   The diagram shows a satellite in circular orbit around the Earth. Which arrow (A, B, C, or D) shows the direction of the resultant force on the satellite?

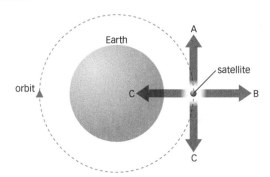

2   A wire in a magnetic field is pushed upwards by the current in it. State **two** different ways of making the push on the wire go downwards instead.

3   Give **two** examples of the use of the motor effect.

4   What does a transformer do?

5   A transformer has more turns on the secondary coil than the primary coil. What type of transformer is it?

6   Describe the magnetic field around a current-carrying straight wire.

7   A ball on a string is whirled around in a circle. How does the tension in the string change when:
    a   the mass of the ball is increased?
    b   the rate of rotation is reduced?

8   A wire carrying a current in the downwards direction passes through a magnetic field that points to the right. In what direction is the wire pushed?

9   Explain why an electric motor has a commutator.

10  Describe the construction of a loudspeaker.

11  A transformer has 1150 turns on its primary coil and 25 turns in the secondary coil. If the primary is connected to the 230 V mains supply, what is the potential difference across the secondary?

12  A current of 2 A at a potential difference of 12 V is drawn from the secondary coil of a transformer. What is the current drawn from the 230 V supply connected across the primary coil?

13  Explain the advantages of using a switch mode transformer to provide power for low-voltage devices.

14  Explain why a transformer can only operate from an a.c. supply.

15  Explain why the coils in the secondary of a step-down transformer need to be of thicker wire than those in the primary.

16  Explain the operation of the brushes and commutator of an electric motor.

**1** The picture shows a racing car going round a corner.

The diagram shows the position of the car, halfway round the corner, viewed from above.

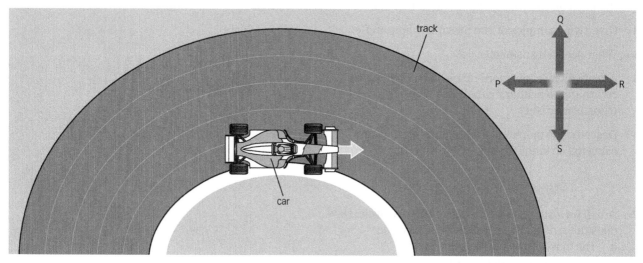

Draw a ring around the correct answer to complete the following sentences.

**a** The | braking / centripetal / gravitational | force on the car is needed for it to turn the corner.

*(1 mark)*

**b** The resultant force on the car is in the direction | P. / Q. / S.

*(1 mark)*

**c** The resultant force on the car comes from | gravity. / friction. / speed.

*(1 mark)*

*(Total marks: 3)*

**2** The diagram shows a transformer designed to operate a 3V light bulb from the mains supply.

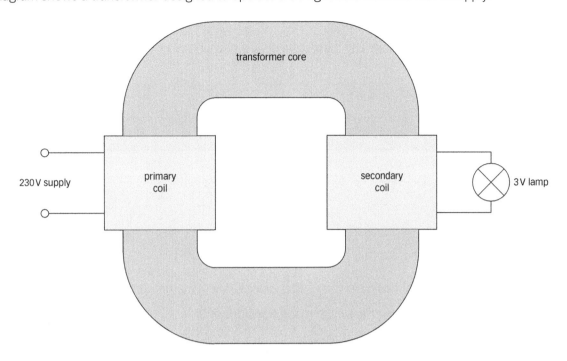

**a** The following sentences explain how the transformer operates.

The sentences are in the wrong order.

**Q** There is a current in the secondary coil.
**R** The current in the primary coil increases.
**S** The magnetic field in the core becomes stronger.
**T** The potential difference across the primary coil increases.
**U** A potential difference is induced across the secondary coil.

Arrange these sentences in the correct order. Start with letter **T**.

$$\boxed{\text{T}} \rightarrow \boxed{\phantom{X}} \rightarrow \boxed{\phantom{X}} \rightarrow \boxed{\phantom{X}} \rightarrow \boxed{\phantom{X}}$$

*(3 marks)*

**b** The secondary coil of the transformer has 90 turns of wire.

Calculate the number of turns of wire in the primary coil.

Write down the equation you use and then show clearly how you work out your answer.

..................................................................................................................................................................

..................................................................................................................................................................

..................................................................................................................................................................

..................................................................................................................................................................

..................................................................................................................................................................

turns of wire = ................................

*(3 marks)*

**c** The current in the lamp is 0.5A.

Calculate the current in the mains supply.

Write down the equation you use and then show clearly how you work out your answer.

..............................................................................................................................................

..............................................................................................................................................

..............................................................................................................................................

..............................................................................................................................................

..............................................................................................................................................

current = ............................... A

*(3 marks)*

**d** A switch mode transformer could also be used to operate the 3V lamp from the mains supply.

Explain the advantages of using the switch mode transformer.

..............................................................................................................................................

..............................................................................................................................................

..............................................................................................................................................

..............................................................................................................................................

..............................................................................................................................................

..............................................................................................................................................

*(4 marks)*

*(Total marks: 13)*

**3** The diagram shows a simple electric motor.

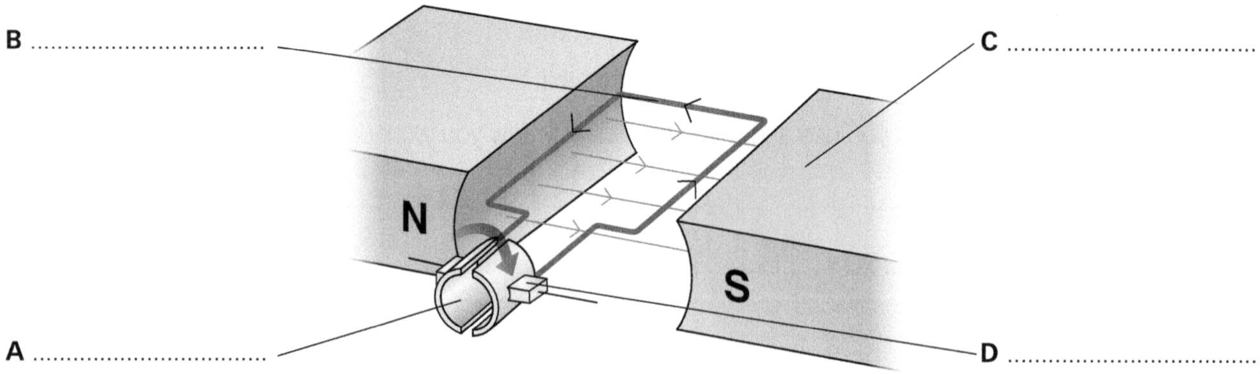

B ....................................

C ....................................

N

S

A ....................................

D ....................................

Use words from the box to label the parts A, B, C, and D.

| battery | brush | coil | commutator | magnet |

*(4 marks)*

*(Total marks: 4)*

## Designing an investigation and making measurements

In this module there are some opportunities to design investigations and make measurements. These include measurements of potential difference and current from transformers, strength of electromagnets, and output forces from electric motors.

Although you may be asked to demonstrate your investigative skills practically, you are also likely to be asked to comment on investigations done by others. The example below offers guidance in this skill area. It also gives you the chance to practise using your skills to answer the sorts of questions that may come up in exams.

## Investigating the strength of electromagnets

Skill – Understanding the experiment

> Sarah tested the hypothesis that the strength of an electromagnet is proportional to the product of the current and the turns of wire in the coil.
>
> She made an electromagnet by coiling some insulated wire around a large iron nail, and connecting it in series with an ammeter to a variable d.c. power supply. For a given number of turns and current, she counted the maximum number of paper clips that the electromagnet could hold up.

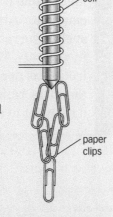

> Sarah's results are in the table.

| Coil current in amperes | Coil turns | Paper clips |
|---|---|---|
| 2.2 | 27 | 6 |
| 1.3 | 15 | 2 |
| 1.7 | 22 | 4 |
| 0.7 | 45 | 3 |

> 1 Identify the **two** independent variables of the investigation.

In an investigation:
- The independent variables are the ones that are changed by the scientist.
- The dependent variable is the one that is measured for each change of the independent variables.

> 2 Suggest some important control variables for the investigation.

A control variable is one that the investigator thinks might affect the outcome, so they try to keep this variable the same all the way through.

Skill – Using data to draw conclusions

> 3 How can Sarah use the data to test her hypothesis?

Sarah needs to use the two independent variables to calculate the independent variable of her hypothesis. Looking for a straight line on a scatter graph of the independent and dependent variables is a good way of establishing proportionality. An alternative is to divide one variable by the other and look for a constant value.

> 4 Explain why Sarah should repeat her experiment more than just four times to get a valid conclusion.

By increasing the amount of data, scientists can spot outliers – data points that might be in error – and double check them. By increasing the range of the independent variable, the hypothesis can be tested more thoroughly. A hypothesis that survives tests over a wide range of variables is more likely to become an accepted theory than one that only predicts results over a narrow range.

Skill – Evaluating the experiment

> 5 Suggest reasons why Sarah might get different results when she repeats the experiment.

There are several reasons why experiments are not exactly repeatable. These include:
- the limited resolution of the measuring instruments
- the effect of uncontrolled variables
- experimenter error, such as misreading an instrument
- systematic errors, such as zero-errors in instruments.

Skill – Societal aspects of scientific evidence

> 6 Describe how Sarah could apply her results to the design of electromagnets for use in scrap yards.

When applying the results of their experiments to make useful devices, scientists need to take account of the effects of changing scale as well as economic factors. A number of prototypes are usually built, starting off at lab scale and getting increasingly larger.

*In this question you will be assessed on using good English, organising information clearly, and using specialist terms where appropriate.*

**1** Explain the operation of a step-up transformer. *(6 marks)*

---

**G–E**

Both coils have to be on a ring of metal. One coil is connected to the mains, this is the prime coil and it magnetises the metal so that electricity can be passed through it to the other coil. More electricity comes out of a step-up transformer than goes in, this is because their is more wire in one coil than in the other.

**Examiner:** This answer earns one mark and is typical of a grade-F candidate. There is very little use of specialist terms, with few details of the required physics. There is no mention of the need for changing magnetism in the core, electricity is used instead of current or potential difference, and length of wire is confused with turns.

---

**D–C**

The potential difference across the primary coil makes field lines which cut the secondary coil. As the field lines are cut, they make a current in the secondary coil. The potential difference has to keep on changing, as magnetoelectric induction can only happen when the magnetic field lines pass through the wires. The power going into the transformer is the same as the power coming out, so its coils of wire don't heat up.

**Examiner:** This answer only earns three marks, and is typical for a weak grade-D candidate. Although specialist terms are spelled correctly, they are often used incorrectly. There was no mention of the iron core, nor was there any discussion of the turns ratio. The last sentence is irrelevant to the question.

---

**B–A\***

A step-up transformer has two coils of insulated wire wound on a thick ring of iron. The primary coil has fewer turns of wire than the secondary coil. The primary coil is connected to a source of alternating current. This creates a magnetic field in the iron, with the field lines pointing along the length of the iron so that they form continuous loops through the secondary coil. Because the field changes direction each time the primary current changes, the magnetic field in the secondary coil keeps on changing. This induces a potential difference across the secondary coil.

**Examiner:** This is a high-quality answer, typical of an A\* candidate, earning six marks. The answer is well organised, with a logical order for the explanation. Correct specialist terms have been used throughout.

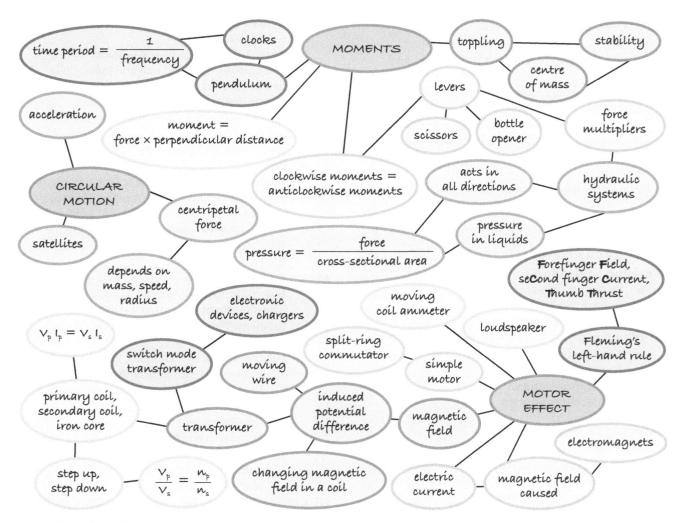

$\text{time period} = \dfrac{1}{\text{frequency}}$ · clocks · MOMENTS · toppling · stability

pendulum · centre of mass

acceleration · levers

$\text{moment} = \text{force} \times \text{perpendicular distance}$ · scissors · bottle opener · force multipliers

CIRCULAR MOTION · clockwise moments = anticlockwise moments · acts in all directions · hydraulic systems

centripetal force

satellites · $\text{pressure} = \dfrac{\text{force}}{\text{cross-sectional area}}$ · pressure in liquids

depends on mass, speed, radius · electronic devices, chargers · moving coil ammeter · Forefinger Field, seCond finger Current, Thumb Thrust

$V_p\,I_p = V_s\,I_s$ · split-ring commutator · loudspeaker · Fleming's left-hand rule

switch mode transformer · moving wire · simple motor

primary coil, secondary coil, iron core · transformer · induced potential difference · magnetic field · MOTOR EFFECT · electromagnets

step up, step down · $\dfrac{V_p}{V_s} = \dfrac{n_p}{n_s}$ · changing magnetic field in a coil · electric current · magnetic field caused

## Revision checklist

- An object's centre of mass is the point at which all of the object's mass seems concentrated.
- The time period of a pendulum is the time taken to complete a full swing.
- Moment is the turning effect of a force.
- When an object is balanced, the total clockwise moment about the pivot is equal to the total anticlockwise moment. This is the principle of moments. Levers multiply forces using moments.
- An object will topple if the line of action of its weight falls outside the base of the object. Stability is increased by making the base wider and the centre of mass lower.
- Liquids are almost incompressible. Pressure in liquids is transmitted equally in all directions.
- Hydraulic systems multiply force.
- An object moving in a circle is constantly changing direction, so is constantly accelerating towards the centre of the circle.
- The force between an object in circular motion and the centre of the circle is a centripetal force.
- Flow of electric current through a wire produces a magnetic field around the wire.

- If a current-carrying wire is placed within another magnetic field, the magnetic fields interact, creating the motor effect.
- When a wire cuts across a magnetic field, a potential difference is induced across the ends of the wire.
- A transformer has an iron core with two coils of wire around it. If the primary coil is connected to an a.c. supply an alternating potential difference is induced in the secondary coil.
- In a step-up transformer, potential difference is higher across the secondary coil than the primary coil.
- In a step-down transformer, potential difference is lower across the secondary coil than the primary coil.
- Potential differences across the primary and secondary coils and the number of turns in the coils are related.
- In a transformer with 100% efficiency, the power input to the primary coil is equal to the output from the secondary coil.
- Most electronic devices use a lower potential difference than the 230 V mains supply. Step-down transformers are used.

## B3 1: Dissolved substances

1  Water
2  Water, carbohydrate/sugar/glucose, ions, and caffeine.
3  Active transport requires carrier protein, osmosis does not.
   Active transport requires energy, osmosis does not.
   Active transport is against a concentration gradient, osmosis is not (or down a concentration gradient).
   Osmosis is the movement of water only.

## B3 1 Levelled questions: Osmosis, sports drinks, and active transport

### Working to Grade E

1  a  Sweating – increases
   b  Temperature – increases
   c  Water intake – increases
2  Diffusion, osmosis, and active transport.
3  a  Sweat more
   b  Become dehydrated
   c  Become more thirsty
4  High water concentration

### Working to Grade C

5  Osmosis is a special kind of diffusion where water moves from an area of high water concentration to an area of low water concentration through a partially permeable membrane.
6  A membrane that allows some molecules through but not others.
7  a  To the left.
   b  Either: water concentration is higher on the right-hand side, or: sugar concentration is higher on the left.
8  Three from: to lubricate joints; to protect organs such as the brain; to carry substances around the body; to help regulate body temperature.
9  Ions
10

| Ingredient | Function of ingredient |
|---|---|
| water | to hydrate the body |
| sugar/carbohydrate/glucose | a source of energy |
| ions | to keep muscles healthy |
| caffeine | to make us more alert |

11  Sports drinks provide sugar/glucose/carbohydrate as a source of energy and ions to keep muscles healthy, whereas water does not contain these.

### Working to Grade A*

12  It is too large to fit through the pores of the partially permeable membrane.

13  a  It increases in size and mass. Cells become turgid (swollen).
    b  There is no change in size or mass.
    c  The chip decreases in size and mass. Cells become flaccid (soft).
14  a  Sports lite
    b  There are less dissolved substances in the drink.
    c  It contains more glucose as an energy source.
    d  Consuming too much sugar could cause a sudden surge in blood sugar levels (a sugar rush), triggering insulin release. This would then be followed by a sudden fall in blood sugar levels (a crash). This sudden change in blood sugar levels would cause problems for the body.
15  a  Some dissolved molecules or ions move from a **low** concentration to a **high** concentration. The movement is **against** a **concentration** gradient. This process is called active transport.
    b  Uptake of ions by root hairs; movement of sodium ions out of nerve cells.
    c  i   Root hairs: nitrates, potassium, phosphates
           Nerves: sodium ions
       ii  Root hairs: soil water
           Nerves: inside the nerve cell cytoplasm
       iii Root hairs: root hair cell cytoplasm and onto the xylem
           Nerves: outside the nerve cell
16  Carrier proteins in the membranes, and a source of energy (ATP).
17  Dead cells cannot make ATP or supply energy, which are necessary to active transport.

## B3 1: Examination questions: Osmosis, sports drinks, and active transport

1  a  The bag would have increased in size and mass. (1)
   b  1 mark awarded for each part of the following answer:
      • There is a concentration gradient between the solutions in the bag and the beaker.
      • Osmosis occurs, causing water to move into the bag of sugar solution (which has a lower water concentration).
      • After 30 minutes, there would therefore be a higher water concentration in the bag of sugar solution.
2  a  1 mark for each of the missing words, up to a total of 4.
      Sports drinks contain carbohydrates, water, and ions. Carbohydrates such as **glucose** are used in respiration to provide **energy**. Water is needed to **hydrate** the body. Mineral ions keep the **muscles** healthy.

**b** 1 mark for identifying both of the best uses.

| Drink contents | Hydration of the body | Energy supply for the body |
| --- | --- | --- |
| Low sugar Dilute drink | ✓ | |
| High sugar Concentrated drink | | ✓ |

**c** During sweating (1)

## B3 2: Exchange surfaces

1 A specialised body surface over which molecules are efficiently exchanged.
2 Digested food such as glucose, amino acids, vitamins, and minerals.
3 Large surface area; thin surface; good blood supply; high turnover.

## B3 3: Gaseous exchange in the lungs

1 To contract and move down to draw air into the lungs, and to relax and move up to expel air from the lungs.
2 Intercostal muscles
3 It diffuses across the alveolus wall, into the blood capillary, and finally into the red blood cell.

## B3 4: Exchange in plants

1 They are the pores in the leaf through which gas exchange occurs. They also control water loss.
2 Any three from: light intensity; humidity; air movement; temperature

## B3 2–4 Levelled questions: Exchange surfaces in the lungs and in plants

### Working to Grade E

1 There is less surface area for diffusion to occur, so diffusion becomes inefficient.
2 **a** To make diffusion more effective.
  **b** Any three from: the digestive system; lungs or gills; plant leaves and roots.
3 In the chest/thorax.
4 The ribs protect the lungs and are used in the process of breathing.
5 Oxygen and carbon dioxide.
6 The ribs move down and in.
7 In the alveoli.
8 The diaphragm
9 The movement of air into and out of the lungs.
10 Mouth → trachea → bronchus → alveolus
11 A – ribs; B – diaphragm; C – lung; D – trachea; E – alveoli
12 Leaves and roots.
13 The movement of water from the roots to the leaves.

14 **a** Guard cells
  **b** **i** Arrow should show movement of oxygen out of the leaf through the stoma.
    **ii** Arrow should show movement of carbon dioxide into the plant through the stoma.
15 Potometer

### Working to Grade C

16 **a**

| Feature | How it improves diffusion |
| --- | --- |
| large surface area | **provides more surface for greater diffusion** |
| thin surface | **provides a short diffusion distance** |
| **efficient blood supply** | to maintain a concentration gradient |

  **b** Any three from: dense capillary network; thin epithelium; presence of a lacteal; large surface area.
17 Surface area to volume ratio is much higher, which allows for efficient diffusion.
18 Any three from: large surface area; thin wall of the alveolus; good blood supply; lining of alveolus is moist.
19 The intercostal muscles relax, causing the ribs to fall down and in, and the diaphragm arches up. Both reduce the volume in the lungs, which increases the pressure.

20

| Gas | Change |
| --- | --- |
| oxygen | decreased |
| carbon dioxide | increased |
| water vapour | increased |
| nitrogen | no change |

21 Any three from: light intensity; temperature; air movement; humidity.
22 **a** Lower
  **b** Less heat from the sun to reduce water loss.
23 **a** The arrow should show water moving out.
  **b** During the day.
24 Larger surface area for exchange of gases. (An examiner would also accept that they can absorb more light.)
25 Increased surface area by the presence of root hairs.
26 It wilts and dies.

### Working to Grade A*

27 Dense capillary network removes absorbed molecules, maintaining a concentration gradient. Thin epithelium provides short distance for molecules to have to move across.
Lacteal removes fats to maintain a concentration gradient.
Large surface area provides greater surface over which molecules can be absorbed.

**28 a**

| Cell dimensions | Surface area | Volume | Surface-area-to-volume |
|---|---|---|---|
| 3 | 54 | 27 | 54:27 2:1 |
| 6 | 216 | 216 | 216:216 1:1 |

   **b** The ratio gets smaller.

   **c** The surface area to volume ratio becomes too small, so diffusion becomes inefficient and it is necessary to have an exchange surface.

**29** The intercostal muscles between the ribs contract, lifting the rib cage up and out, expanding the thorax.
The diaphragm contracts and flattens. This expands the thorax.
The volume inside the lungs increases, and the pressure decreases.
Air rushes into the lungs due to the low pressure.

**30** To reduce water loss.

**31 a** Increasing the light will increase the rate of transpiration.

   **b** Increasing the air movement will increase the rate of transpiration.

   **c** Increasing the humidity will decrease the rate of transpiration.

## B3 2–4 Examination questions: Exchange surfaces in the lungs and in plants

**1 a** 1 mark is awarded for the working, and 1 mark for the correct answer.
$320 \div 20 = 16$ breaths per minute

   **b** It will increase. (1)

   **c** **i** Oxygen (1)
      **ii** 1 mark awarded for each of the following points, up to a total of 3:
- it has a large surface area
- it has a good blood supply
- it has a thin wall – only one cell thick.

**2 a** As the light intensity increases, the rate of water loss increases, but eventually there will be no further increase in water loss despite an increase in light intensity. (1)

   **b** **i** They close. (1)
      **ii** The rate of transpiration decreases with lower light intensity. (1)

   **c** 2 marks awarded for explaining that the graph shows that high light levels (1) and windy conditions (1) increase the rate of transpiration (water loss), so the gardener will have to water their plants most on sunny and windy days.

## B3 5: Circulation and the heart

**1** To transport substances such as glucose and oxygen around the body.

**2** They provide a short-term solution to heart failure that keeps patients alive while they await a heart transplant.

**3** The muscle walls of the heart contracts. This forces blood through and out of the heart. The blood is pumped into arteries, which carry the blood around the body. Any backflow is prevented by valves.

## B3 6: The blood and the vessels

**1** Away from the heart.

**2** It carries blood under higher pressure than in a vein.

**3** White blood cells fight infection. Some engulf and digest microorganisms; others make antibodies that kill microorganisms. Platelets help to form blood clots, which prevent microorganisms entering the body.

## B3 7: Transport in plants

**1** In the xylem by the transpiration stream.

**2** In the phloem by translocation.

**3** The source (the leaf) to the sink (rest of plant, such as the root, seed, etc.).

## B3 5–7 Levelled questions: Transport in animals and plants

### Working to Grade E

**1** Blood, heart, and blood vessels.

**2** Oxygen

**3 a** A – right atrium; B – left atrium; C – left ventricle; D – right ventricle

   **b** Blood vessel leaving left ventricle.

   **c** Any of the valves between the atria and ventricles, or between the ventricles and the arteries.

**4** Artery

**5** Vein

**6** Any two from: carbon dioxide, soluble products of digested foods, and urea.

**7 a** To transport oxygen around the body.

   **b** Haemoglobin

**8** Water and mineral ions.

**9** Dissolved sugars

### Working to Grade C

**10** The blood passes through the heart twice per cycle around the body.

**11 a** Pulmonary artery

   **b** Vena cava

**12** Arrows should show blood flow from upper and lower vena cava into right atrium, down through valve into right ventricle, then up through valve and out via left and right pulmonary arteries.

**13** To prevent the backflow of blood.

**14 a** The ventricles need thicker walls to produce a more powerful contraction, since they have to pump the blood out of the heart, while the atria only pump blood to the ventricles.

   **b** It has to pump blood all round the body, while the left ventricle pumps blood to the lungs only.

**15** They would fit an artificial valve.

16 a Any one from: keeps the patient alive; it is not rejected by the body.
   b Any one from: it is a short-term solution; they often have wires that protrude through the skin.
17 a White blood cell.
   b It helps form blood clots.
   c It creates more room for haemoglobin.
18 Transport, protection, and regulation.
19 Any time when real blood is not available, for example during a war or major trauma.
20 A device to widen a blood vessel.
21 Either: xylem has dead cells, phloem has living cells, or: xylem has large angular cells, phloem has smaller cells.
22 The movement of dissolved sugars through the phloem from the leaves/site of photosynthesis (source) to the rest of the plant (sink), including growing regions and storage organs.
23 a A site of photosynthesis, where sugars are made.
   b Leaf
   c Growing regions and storage organs.

## Working to Grade A*

24 a Any three from: arteries have a thicker wall than a vein; arteries have larger amounts of muscle; arteries have larger amounts of elastic fibres; arteries have narrower lumens; veins have valves while arteries do not.
   b Arteries have a thicker wall than a vein because blood is under high pressure in the artery so a thick wall withstands the pressure. Arteries have larger amounts of muscle than veins as it allows the wall to withstand and maintain the high pressure.
   Arteries have larger amounts of fibres than veins to allow stretch and recoil.
   Arteries have narrower lumens than veins to help maintain pressure.
   Veins have valves to prevent backflow of blood.
25 To allow substances to diffuse through the capillary wall quickly.
26 It replaces lost volume, and will not be rejected by the body.
27 The stent/a wire mesh is inserted into the narrowed region of the artery. The stent is opened/expanded using a small balloon, and this opens the artery to ensure proper blood flow.
28 Water molecules move into the root hair cell. They move from cell to cell across the root into the xylem. They travel up the xylem to the stem and leaves. From the leaf cells, water molecules move into the air spaces in the leaf, then out through the stoma as water vapour.

## B3 5–7 Examination questions: Transport in animals and plants

1 a To carry oxygen around the body. (1)
  b They form a clot at the site of a cut (a scab), which prevents microorganisms entering the body. (1)

  c Phagocytes ingest/digest microorganisms. Lymphocytes make antibodies to kill microorganisms. 1 mark for identifying each type up to a total of 2.
2 a They prevent the backflow of blood. (1)
  b 2 marks awarded for two different advantages; 2 marks awarded for two different disadvantages, up to a total of 4. For example:
  **Advantages:** they extend lifespan; they improve quality of life.
  **Disadvantages:** surgery carries with it a risk of infection; the valve can damage blood cells and this results in blood clots, which are a cause of heart attacks/strokes; the patient will have a lifelong dependency on anticlotting drugs, which are expensive.
3 a Pulmonary artery (1)
  b The lungs (1)
  c 8 marks available. Any of these points would be acceptable and earn 1 mark each (up to a total of 8) as long as the full journey of the blood from the vena cava to the aorta is explained.
  • Blood from the vena cava enters the right atrium.
  • From the right atrium, the blood enters the right ventricle.
  • Blood passes through the open valve.
  • The right ventricle contracts, forcing the blood out.
  • Blood enters the pulmonary artery.
  • Blood travels to the lungs.
  • Backflow of blood is prevented by the valve.
  • Blood returns from the lungs in the pulmonary vein.
  • Blood enters the left atrium.
  • Blood passes into the left ventricle.
  • The left ventricle contracts, forcing blood out into the aorta.
  References to valves on both left and right sides will not be credited.

## How Science Works: Exchange and transport

1 a i 1 mark each for diaphragm and intercostal muscles.
    ii **Similarities:** 1 mark for either that they expand the chest or they cause air to be drawn into the lungs (ventilation).
    **Differences:** 1 mark for explaining that muscles cause the chest to expand in the body, but changes in pressure carry out this function in the iron lung.
  b 1 mark for each answer given, up to a total of 3.
  **Advantages:** they keep patients alive while they recover (from diseases which paralyse the muscles).
  **Disadvantages:** they are expensive; they are difficult for doctors to treat patients within; patients have a poor quality of life.

2 a 1 mark for each point up to a maximum of 3. They allow more freedom of movement/ improve quality of life; they allow the doctors access to the body for examination and surgery; they can be used during open chest surgery when the ribs don't function.

## B3 8: Homeostasis and the kidney

1 To maintain the normal functioning of the body cells.

2 Body temperature, blood glucose/sugar levels, water content, ion (salt) content, and pH.

3 Blood containing waste products arrives at the kidney; the blood is filtered in the outer zone of filtration; useful substances and water are re-absorbed in the inner zone of re-absorption; filtered blood leaves the kidney and waste passes down to the bladder in the urine.

## B3 9: Kidney treatments

1 Replacing a damaged kidney with one from another person.

2 Blood is passed into a (dialysis) machine. Waste products are filtered out and the concentrations of all dissolved substances in the blood, such as salts, are restored to normal.

3 Antigens on the surface of the new organ are not recognised by the body. The body produces antibodies to attack the 'alien' organ and this leads to organ rejection.

## B3 10: Controlling body temperature

1 Any one from: hairs flatten; sweat is released; vasodilation.

2 The thermoregulatory centre (hypothalamus) is an area of the brain. It monitors the temperature of blood flowing through the brain and receives messages from nerves about skin temperature. If temperature is too high or too low, it triggers the body's response to temperature change.

3 This is where the blood vessels in the surface of the skin expand or dilate. This causes more blood to flow to the surface of the skin. It results in greater heat being lost by radiation.

## B3 11: Controlling blood sugar

1 Insulin and glucagon.

2 Insulin

3 Low blood glucose levels are detected by the pancreas. It stops making insulin, and starts to make glucagon. The glycogen in the liver is broken down into glucose. The glucose is released into the blood, restoring the blood glucose level to normal.

## B3 8–11 Levelled questions: Homeostasis
### Working to Grade E

1 Homeostasis is the maintaining of a constant internal environment.

2 a Temperature control
  b Water and ion control
  c Blood sugar control

3 The removal of waste products made in the body.

4 Carbon dioxide and urea.

5 37 °C

6 To remove wastes such as urea, excess water, and ions.

7 Abdomen

8 Dialysis and transplant.

9 Any two from: drugs; disease; diabetes; genetic causes.

10 Thermoregulation

11 Being in an environment with a high external temperature; exercise; dehydration.

12 When the body temperature drops below 35 °C.

13 It is an area of the brain/the hypothalamus.

14 Diet

15 Any one from: genetic; viral; some drugs; trauma.

16 As a source of energy.

17 Mainly in the liver.

18 It decreases blood glucose levels.

### Working to Grade C

19 a Filtration of the blood. Waste products such as urea and small useful substances are removed.
   b The bladder

20 Carbon dioxide or lactic acid produced during respiration.

21 Produced when excess amino acids from proteins are broken down in the liver.

22 They go down.

23 Antigens are protein markers on the surface of cells.

24 The donor is the person who gives the kidney, the recipient is the person who receives the kidney.

25 a Arrow A should show movement from patient's arm, along line artery to vein pump, into tubes in dialysing solution, and out into container of used dialysing solution.
   b Label should be to the tubes inside the dialysis machine which are surrounded by dialysis fluid.
   c i 5–6 hours
     ii Wastes continually build up in the body and the patient would quickly get very ill if the toxins were not removed.

26 a Vasodilation: blood vessels in the surface of the skin expand. More blood flows to the skin.
   b More heat is lost from the skin by radiation.

27 a A hot day.
   b Any three from: hairs flattened; vasodilation/ blood vessels enlarged; sweat production; radiated heat from blood.

28 Nerves in the skin detect skin temperature.

29 The temperature denatures enzymes and this will harm cells.

30 Checks on circulation and eyesight.

31 Blood becomes more concentrated; water is withdrawn from cells by osmosis. Over the long-term, raised blood glucose is very damaging to a person's health.

32 The pancreas

33 Young people

34 No injections are needed and the patient can lead a more normal life.

35 Thirst and frequent urination.

36 a i After 20 minutes/around 4.20.
    ii This is when the level of blood glucose concentration starts to rise.
  b i John
    ii His glucose level is higher at the start. It rises higher, and is slower to return to normal.
  c i No change.
    ii Will rise after the meal, and then fall as the level of blood glucose returns to normal.
  d Inject insulin.

## Working to Grade A*

37 The zone of re-absorption. Here, useful molecules (sugar, dissolved ions, water to ensure correct hydration) are reabsorbed by active transport.

38 It will affect osmosis, causing water to move into or out of cells, resulting in damage.

39 The breathing rate increases to remove carbon dioxide from the blood (via gaseous exchange in the lungs) more quickly.

40 a To maintain a concentration gradient in order to encourage diffusion of wastes out of the blood.
  b The dialysis fluid contains the correct levels of salts, and so there is no diffusion of salts out of the blood.

41 Acute failure is when the kidneys stop working suddenly.
Chronic failure is when the kidneys gradually fail to work.

42 a When antibodies are produced by the immune system which attack and damage the new kidney.
  b Tissue typing prior to transplant; use of immunosuppressant drugs.

43 Dialysis limits the patient's quality of life, while a transplant will improve their quality of life. Transplants are cheaper in the long run, and a longer term solution.

44 When the body temperature rises, sweat is released from the sweat glands and covers the surface of the skin. Body heat is used to evaporate the sweat, and this heat loss cools the body and helps to regulate body temperature.

45 On a hot day, we sweat more to cool the body down. This means we need to drink more to replace lost water/ensure proper hydration.

46 When hairs stand up, they trap more insulating air and this keeps us warm. When hairs lie flat, they trap less air, reducing insulation. This cools the body.

47 During vasoconstriction, blood vessels in the surface of the skin get narrower. This means less blood flows near the surface of the skin, so less heat is lost by radiation. This helps to warm the body and maintain core body temperature.

48 When blood glucose levels becomes low, the pancreas detects the fall. It releases the hormone glucagon into the blood. Glucagon causes the liver to release glucose into the blood, which causes the blood glucose level to return to normal.

49 Modern sensors are simple and more effective, making it easier for diabetics to keep track of blood glucose levels.
Genetically engineered human insulin poses less risk of allergies, making it safer.
Thorough checking of circulation reduces complications that could cause serious illness.
Automated insulin pumps allow patients to lead a more normal life.

## B3 8–11 Examination questions: Homeostasis

1 a $2650\,cm^3$ (1)
  b Sweating (1)
  c Vasodilation/hairs lie flat. (1)

2 a They might have eaten a sugary meal. (1)
  b David (1)
  His blood glucose level was higher before the meal. (1)
  When the level rises after eating a meal, it does not return to normal quickly. (1)
  c i Insulin (1)
    ii It would increase. (1)
    iii The pancreas (1)
    iv The liver (1)
    v Glycogen (1)
  d Their pancreas fails to produce enough insulin. (1)

3 a Urea (1)
  b 1 mark for each point, up to a total of 4:
    • When the body is dehydrated,
    • the kidney produces little and concentrated urine.
    • When the body is over hydrated,
    • the kidney produces large amounts of dilute urine.
  c i It is toxic to cells because it is alkaline. (1)
    ii Any three from the following list, earning 1 mark each up to a maximum of 3:
      • There will be no change in glucose, sodium and chloride ions
      • because the concentrations are the same in the blood and the fluid.
      • Urea and potassium ions will diffuse out
      • because there is a higher concentration in the blood than in the dialysis fluid.

**d** 6 marks available. 3 marks will be awarded for three different causes, and 3 marks will be awarded for a sufficient explanation of how to overcome the problems identified.
Rejection is caused by:
- antigens on the surface of kidney cells
- are not recognised as part of the body
- antibodies are produced by the immune system
- antibodies attack the alien kidney.

To prevent rejection:
- tissue typing is carried out
- to match the donor antigens with the recipient's
- antigens on the kidney are recognised as part of the body
- no antibodies are produced
- immunosuppressant drugs are used.

## B3 12: Human populations and pollution

1  The larger the population, the greater the level of pollution.
2  Carbon dioxide, sulfur dioxide, smoke.
3  It removes food sources and shelter for many species. They can no longer survive. This reduces the biodiversity/number and types of organisms in the area.

## B3 13: Deforestation

1  Any one from: timber for building, furniture, and fuel; to clear land to grow crops, e.g. biofuels, cash crops, and feed livestock, and to build farms, towns, and industries.
2  It reduces biodiversity.
3  Deforestation increases global levels of carbon dioxide, because there is less photosynthesis so less carbon dioxide is taken up by plants, and more decay and burning, both of which release carbon dioxide.

## B3 14: Global warming and biofuels

1  Biogas, wood, or alcohol.
2  It releases greenhouse gases such as carbon dioxide. These trap heat in the atmosphere.
3  It causes loss of habitats, such as the melting of ice caps. This forces animals to change distribution and migration patterns. Global warming may reduce biodiversity – some species will die out as their habitats are lost. Climate changes such as droughts are stressful for animals.

## B3 15: Food production

1  It is a protein-based food product made from a fungus. It is used as a food source in meat substitutes.
2  They have led to overfishing, resulting in falling fish stocks in the oceans.

## B3 12–15 Levelled questions: Humans and their environment

### Working to Grade E
1  The number of individuals of a species in an area.
2  **a**  10 million
   **b**  Approximately 150 years.
3  Large scale felling of trees.
4  Used to make buildings/furniture or as a fuel.
5  **a**  Nutrient-rich composts
   **b**  Releases carbon dioxide as it decays.
   **c**  Use peat-free compost.
6  The overall increase in average global temperatures.
7  A range of fuels made from biological materials.
8  Carbon dioxide and methane.
9  Methane
10  **a**  Dead plant and animal waste.
    **b**  It supplies fuel for cooking.
11  Any three from: habitat loss; species distribution changes; ice caps melt; climate change; migration patterns change.
12  A large tank for microorganisms to be grown in.
13  The number of fish it is permitted for a country to catch.

### Working to Grade C
14  Humans reduce the amount of land available for other species and release pollutants.
15  **a**  High mortality rate, especially among infants, due to poor hygiene and healthcare. Low numbers of individuals from which to breed. Food supply was limited and people ate a poor diet.
    **b**  **i**  20th century
         **ii**  Improved diet, hygiene, and healthcare.

16

| Pollutant | Source | Effect on the environment |
|---|---|---|
| smoke | released from burning fossil fuels | **causes bronchitis, and reduces photosynthesis** |
| **carbon dioxide** | **released from burning fossil fuels** | contributes to global warming, and acid rain |
| sulfur dioxide | **released from burning fossil fuels** | **forms acid rain** |

17  **a**  Using resources without harming the environment.
    **b**  Avoid overuse of resources; handle waste correctly; recycle materials; replace resources where possible.
18  Any three from: building towns and industrial areas; quarrying; farming; landfill waste sites.
19  Carbon dioxide is released during burning; carbon dioxide is released during the decay of felled trees; there is a reduction in photosynthesis because there are fewer trees to take up carbon dioxide.

**20** **a** $100 \div 0.8 = 125$ years

**b** Land is being used for farming, allowing cash crops to be grown for rapid income. Production of bioethanol.

**21** Carbon dioxide from burning fossil fuels and deforestation; methane from cattle, rice fields, and decaying waste.

**22** **a** Anaerobic fermentation of carbohydrates in plant material and sewage by bacteria.

**b** It supplies energy in remote areas, where national grids cannot reach.

**23** It reduces biodiversity by contributing to climate change that causes habitat loss.

**24** Glucose

**25** Less movement of animals, so less energy used. Less energy lost as heat, as being close together the surroundings become warmer.

**26** Fish stocks have declined due to overfishing.

**27** At every link in the food chain, more energy is lost. The longer the food chain, the more energy is lost. Therefore eating producers means less energy has been lost.

**28** pH and temperature

**29** Sonar and efficient, sophisticated nets.

**30** There has been a huge increase in the human population, so we need to provide enough protein to feed everyone.

## Working to Grade A*

**31** It removes food and shelter for many species. It reduces biodiversity. It could cause many species to die out (become extinct).

**32** **a** They contain nitrates, which kill fish by a process called eutrophication.

**b** They build up in food chains to toxic levels.

**33** Untreated sewage contains bacteria that can cause disease and nitrates, which can kill fish by eutropication.

**34** **a** The greater the deforestation, the greater the loss of forest habitats, and the greater the reduction in biodiversity.

**b** There is reforestation.

**35** In a large-scale generator, waste is constantly added and much larger volumes of gas are produced.

**36** They absorb large amounts of carbon dioxide as it dissolves in the water. Phytoplankton absorbs carbon dioxide during photosynthesis.

**37** The plants take in carbon dioxide to grow and release it again when burnt. So there is no overall increase in carbon dioxide levels.

**38** In battery farming, the chickens are reared in small cages, whereas in free range farming, the animals are allowed to roam freely.

**39** It allows smaller/younger fish to escape so they can survive and breed.

**40** The greater the amount of transport involved, the more fuel is burnt. This releases pollutant gases such as carbon dioxide into the atmosphere, which contributes to global warming.

**41**

| Pros | Cons |
|---|---|
| Less energy is lost in the food chain, so more is available for human consumption. | Greater risk of disease spreading through the animals as they are in close contact. |
| Less labour intensive, as animals are all contained in a limited area. | Some people feel that the technique is inhumane, or cruel to the animals. |
| Less risk of attack from predators such as foxes. | Some people believe that the quality of the product is poorer. |
| Production costs are cheaper. | |

In summary, the pros lead to cheaper, more plentiful food. The cons are inhumane treatment of animals, and the quality of the food is not as good.

## B3 12–15 Examination questions: Humans and their environment

**1** 1 mark will be awarded for each correct match, up to a maximum of 4.

| Burning fossil fuels releasing sulfur dioxide. | Dissolves in rain to form acid rain. |
|---|---|
| Pesticides are used by farmers to kill pests. | Builds up in the food chain, killing other organisms. |
| Release of sewage. | Causes the death of fish by eutrophication. |
| Growing large areas of rice which release methane. | Contributes to global warming. |

**2** **a** Lack of movement – uses less energy. (1)
Birds lose less energy as heat – if they are close together they can share body heat. (1)

**b** Any three from the following list earning 1 mark each up to a total of 3:
- it causes behavioural problems such as scratching
- diseases spread rapidly through the chickens, as they are so close together
- there is an increased use of antibiotics
- there are ethical issues about the humane treatment of animals.

**3** **a** It can become explosive. (1)

**b** It keeps it warm. (1)

**c** Any two from the following list, earning 1 mark each up to a total of 2:
- it provides fuel for cooking and heating
- rural communities cannot always be on mains gas supply
- it disposes of waste
- it reduces use of fossil fuels.

## C3 1: The periodic table

1   Newlands listed the elements then known in order of atomic weight. Every eighth element had similar properties. He used this pattern to group the elements. He called his idea the 'law of octaves.'

2   Mendeleev swapped the positions of some pairs of elements so that they were grouped with elements with similar properties; Mendeleev left gaps for elements which he predicted did exist, but which had not then been discovered.

3   Strong, hard, high density, high melting point, form ions with different charges, form coloured compounds, react slowly or not at all with water and oxygen.

## C3 2: Group 1 – The alkali metals

1   Three rows from:

| | Group 1 elements | Transition elements |
|---|---|---|
| Density | low | high |
| Hardness | soft | hard |
| Strength | strong | weak |
| Melting point | low | high (except mercury) |
| Compound colour | white | coloured |
| Reaction with oxygen | vigorous | not vigorous, or does not occur |
| Reaction with water | vigorous | not vigorous, or does not occur |

2   Similarities: Both lithium and potassium react with water to form a hydroxide and hydrogen gas. The reactions are vigorous, and the hydrogen gas propels the metal around on the surface of the water as it reacts.
Differences: a lilac flame is seen when potassium reacts with water. There is no flame when lithium reacts with water.

3   The trend can be explained by the energy level of the outer electrons. Potassium is lower down Group 1 than lithium. The outermost electron of potassium is in a higher energy level than that of lithium. This means that, in reactions, potassium gives away its outermost electron more easily than lithium. Potassium is more reactive than lithium.

## C3 3: Group 7 – The halogens

1   Going down the group, the melting points and boiling points increase.

2   Iron and chlorine have the more vigorous reaction, because chlorine is more reactive than iodine.

3   a

| chlorine | + | sodium iodide | → | sodium chloride | + | iodine |
|---|---|---|---|---|---|---|
| $Cl_2$ | + | $2NaI$ | → | $2NaCl$ | + | $I_2$ |

b

| bromine | + | potassium iodide | → | potassium bromide | + | iodine |
|---|---|---|---|---|---|---|
| $Br_2$ | + | $2KI$ | → | $2KBr$ | + | $I_2$ |

## C3 1–3 Levelled questions: The periodic table

### Working to Grade E

1   transition; different; coloured; catalysts

2   a   1
    b   3
    c   7

3   a   In the periodic table, a vertical column is called a group.
    b   True
    c   Mendeleev left gaps in his periodic table for elements he predicted did exist, but had not yet been discovered.

4   a   manganese
    b   potassium
    c   iron
    d   manganese

### Working to Grade C

5

| Name of compound | Formula of metal ion in compound | Appearance and state of compound at room temperature |
|---|---|---|
| potassium chloride | K+ | white solid |
| sodium bromide | Na+ | white solid |
| lithium chloride | Li+ | white solid |

6   a   sodium + chlorine → sodium chloride
    b   lithium + oxygen → lithium oxide
    c   sodium + water → sodium hydroxide + hydrogen
    d   potassium + water → potassium hydroxide + hydrogen

7   Equations for the pairs that react:
    a   chlorine + potassium bromide → potassium chloride + bromine
    b   No reaction
    c   bromine + potassium iodide → potassium bromide + iodine
    d   No reaction
    e   chlorine + potassium iodide → potassium chloride + iodine

8   a   1
    b   7
    c   1
    d   7
    e   1
    f   1

## Working to Grade A*

**9**　**a**　$2Na(s) + Br_2(l) \rightarrow 2NaBr(s)$

　　**b**　$2Li(s) + 2H_2O(l) \rightarrow 2LiOH(aq) + H_2(g)$

　　**c**　$Cl_2 + 2KI \rightarrow 2KCl + I_2$

**10**　When a metal reacts with a halogen, the metal atoms give each halogen atom **one** extra electron. This completes the **outer** energy level of the halogen atom. The closer the outer energy level is to the nucleus, the **greater** the attraction between the newly-added electrons and the nucleus. So the lower down Group 7 an electron is, the **less** easily its atoms gain electrons, and the less reactive the element is.

## C3 1–3 Examination questions: The periodic table

**1**　**a**　**i**　Mendeleev used the atomic weights published by Cannizzaro to arrange the elements in order. (1)

　　　**ii**　Mendeleev's predictions, of the positions of the elements in the periodic table that had not then been discovered, were correct. (1)

　　**b**　**i**　1 mark awarded for either atomic number/proton number.

　　　**ii**　3 marks available: 1 mark will be awarded for your explanation, 1 mark for including examples and 1 mark for providing at least one correct electronic structure of an element. For example:
All the elements in a group have the same number of electrons in their highest occupied energy level. The number of electrons in the highest occupied energy level is the same as the group number, for the main groups. For example the electronic structure of sodium (a member of Group 1) is 2.8.1. This shows it has 1 electron in its highest occupied energy level.

**2**　**Here is a guide to the marking for an essay-style question:**
To gain 5/6 marks – All information in the answer must be relevant, clear, organised, and presented in a structured and coherent format. Specialist terms should be used appropriately. Few, if any, errors in grammar, punctuation, and spelling. Answer should include 5 or 6 points from those below.
To gain 3/4 marks – Most of the information should be relevant and presented in a structured and coherent format. Specialist terms should usually be used correctly. There may be occasional errors in grammar, punctuation, and spelling. Answer should include 3 or 4 points from those below.
To gain 1/2 marks – Answer will be simplistic. There may be limited use of specialist terms. Errors of grammar, punctuation, and spelling will prevent communication of the science. Answer should include 1 or 2 points of those listed below.

*Points to include:*

- Transition elements (except mercury) have higher melting points than the Group 1 elements.
- Transition elements have lower densities.
- Transition elements are stronger.
- Transition elements are harder.
- Transition elements form coloured compounds, Group 1 elements form white compounds.
- Transition elements form ions with different charges, Group 1 elements form ions with a charge of +1.
- The Group 1 elements react vigorously with oxygen and water, the Transition elements react less vigorously, if at all.

**3**　**a**　Gas (1)

　　**b**　The boiling point increases. (1)

　　**c**　hydrogen + fluorine → hydrogen fluoride (1)

　　**d**　**i**　Iodine (1)

　　　**ii**　2 marks available. 1 mark will be awarded for naming the product and 1 mark for explaining how they will react. For example: The element will react with hydrogen to form hydrogen iodide. The reaction will be less vigorous than the reaction of hydrogen with bromine.

　　**e**　**i**　A maximum of 4 marks are available. 1 mark will be awarded for describing the trend in reactivity. Then 1 mark per reaction observation will be awarded up to a maximum of 2. 1 mark per displacement reaction observation will be awarded up to a maximum of 2. For example:
Going down the group, the elements get less reactive. The reactions of the halogens with hydrogen illustrate this trend – the reaction with fluorine happens spontaneously and explosively; the reaction of hydrogen with bromine requires a lighted splint to get it started.
In group 7, a more reactive halogen displaces a less reactive halogen from an aqueous solution of its salt. For example, chlorine displaces iodine from a solution of sodium iodide because chlorine is more reactive than iodine. Iodine does not displace chlorine from sodium chloride solution because iodine is les reactive than chlorine.

　　　**ii**　This trend can be explained by the energy levels of the outer electrons in halogen atoms. (1) For example, when hydrogen reacts with chlorine, hydrogen atoms give each chlorine atom one extra electron. The electron completes the outer energy level of the chlorine atom, forming a chloride ion. The higher the energy level of the outer electrons, the less easily electrons are gained. (1)

## C3 4: Making water safe to drink

1   A suitable source is chosen; the water is filtered; the water is sterilised by adding chlorine.

2   Advantages – prevents tooth decay; less money spent treating dental problems.
    Disadvantages – expensive; would be unnecessary if everyone looked after their teeth properly; very large amounts of fluoride compounds can make teeth yellow.
    A conclusion for or against adding fluorine to water should then be added, and backed up by a reason.

3

| Type of water filter | What it removes |
|---|---|
| carbon | chlorine and other molecules with unpleasant tastes and smells |
| silver | bacteria |
| ion exchange resin | cadmium, lead, copper ions (and calcium and magnesium ions) |

## C3 5: Hard and soft water

1   Dissolved calcium and magnesium ions make water hard.

2   Hard water advantages – calcium ions prevent heart disease and help the development and maintenance of teeth and bones.
    Hard water disadvantages – forms scale in kettles and boilers, increasing energy bills and maybe increasing emission of greenhouse gases; forms scum with soap, so increasing amount of soap needed.
    Soft water advantages – does not form scale or scum.
    Soft water disadvantages – no calcium ions for maintenance of teeth and bones.

3   Boiling – this removes temporary hardness; adding sodium carbonate – this removes both types of hardness; passing water through an ion exchange resin – the calcium and magnesium ions are replaced by sodium, or hydrogen, or potassium ions.

## C3 4–5 Levelled questions: Water

### Working to Grade E

1   a, c, d, e = H; b = S

2   Choose an appropriate source – to minimise the treatment needed to make the water safe to drink;
    Pass the water through filter beds – to remove solids from the water;
    Add chlorine – to kill bacteria in the water.

3   lather; scum; more; detergents

4   Compounds that may be dissolved in hard water: a, c, d

### Working to Grade C

5   Arguments for: prevents tooth decay, reduces dental treatment bills.
    Arguments against: expensive, unnecessary for those who look after their teeth properly.

6   Permanent hardness is not removed on boiling; temporary hardness is removed on boiling.

7   a   Anomalous result – village C, run 3
    b   A – 3; B – 20; C – 44
    c   So that she can calculate a mean from the three results so increasing the accuracy of her data **or** to identify anomalous results.
    d   A, B, C
    e   A
    f   B and C

8   1 – D; 2 – A; 3 – F and B; 4 – C; 5 – E

9   Sodium carbonate is soluble in water. In hard water, its carbonate ions react with dissolved calcium and magnesium ions. Calcium carbonate and magnesium carbonate form as precipitates. They can be removed by filtering.

10   The energy to evaporate the water is not supplied from the Sun, but is supplied from fossil fuel sources, for example.

### Working to Grade A*

11   D, A, E, C, F, B

## C3 4–5 Examination questions: Water

1   **Here is a guide to the marking for an essay-style question:**
    To gain 5/6 marks – All information in the answer should be relevant, clear, organised, and presented in a structured and coherent format. Specialist terms will be used appropriately. Few, if any, errors in grammar, punctuation, and spelling. Answer should include 5 or 6 points from those below.
    To gain 3/4 marks – Most of the information should be relevant and presented in a structured and coherent format. Specialist terms will usually be used correctly. There may be occasional errors in grammar, punctuation, and spelling. Answer should include 3 or 4 points of those listed below.
    To gain 1/2 marks – Answer will be simplistic. There may be limited use of specialist terms. Errors of grammar, punctuation, and spelling will prevent communication of the science. Answer should include 1 or 2 points of those listed below.
    *Points to include:*
    • Chlorine kills bacteria in water from source to tap.
    • Adding chlorine to water prevents illness and death from waterborne diseases.
    • Adding chlorine to water reduces health bills.
    • Some people do not like the taste and smell of chlorine in tap water.
    • Adding chlorine to tap water means that companies can make money selling water filters that remove it from water.
    • Adding chlorine to water increases the costs of water treatment, compared to not adding it.
    • Conclusion, backed up by reasons.

**2 a** The student has a water softener at home, but it doesn't work when the column is saturated with calcium ions (1);
On different days, the water company supplies water from different sources (1).

**b** 1 mark will be awarded for each of the following points, up to a maximum of 3:
- Add soap solution from the burette to each water sample in turn.
- Record the volume of soap solution required to form permanent lather.
- The smaller the volume required, the softer the water.

**c i** 2 October (1)

**ii** Take more than one sample on each date, find the volumes of soap solution required for each one, and calculate the mean value added. (1)

**iii** 2 October (1)

**3 a i** calcium (1) and magnesium (1).

**ii** The water in beaker A has more $Ca^{2+}$ and $Mg^{2+}$ ions in it. (1)
The water in beaker B has more $Na^+$ ions in it. (1)

**b** 3 marks available. 1 mark awarded for an advantage and 1 mark for a disadvantage, up to a total of 2.
Advantages – less scum with soap, and less scale in kettles, reducing bills for soap and energy.
Disadvantages – less calcium in water to help maintain healthy bones and teeth, less calcium to help protect against heart attacks.
Final mark awarded for a conclusion, with supporting evidence.

**4 a** Temporary hard water is softened on boiling, permanent hard water is not softened on boiling. (1)

**b** Sodium carbonate is soluble in water. In hard water, its carbonate ions react with dissolved calcium and magnesium ions. (1) Calcium carbonate and magnesium carbonate form as precipitates. They can be removed by filtering. (1)

**c** Boil the water. (1) Temporary hard water contains hydrogen carbonate ions ($HCO_3^-$). On heating, these ions decompose to produce carbonate ions ($CO_3^{2-}$). (1) The carbonate ions react with calcium ions ($Ca^{2+}$) or magnesium ions ($Mg^{2+}$) in the water to make calcium carbonate or magnesium carbonate. (1) Calcium carbonate and magnesium carbonate are insoluble in water, so they form as precipitates. The precipitates form scale in kettles and boilers. The boiled water contains few calcium or magnesium ions. It has been softened. (1)

## C3 6: Calculating and explaining energy change

1 Joules, J
2 29 400 J
3 To help minimise heat losses to the surroundings.

## C3 7: Energy-level diagrams

1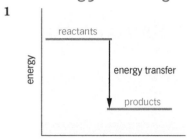

2 A catalyst provides a different pathway for a chemical reaction that has a lower activation energy.
3 The energy released in forming four new O–H bonds in the water molecules is greater than the energy needed to break the existing two H–H and one O=O bonds.

## C3 8: Fuels

1 Hydrogen can be burned as a fuel in combustion engines, forming water.
It can be also used in fuel cells that produce electricity to power vehicles.
2 Carbon dioxide (a greenhouse gas) is produced; many hydrocarbon fuels are not renewable; burning hydrocarbons at high temperatures makes oxides of nitrogen, which destroy ozone in the upper atmosphere. **There are many other possible correct answers.**

## C3 6–8 Levelled questions: Calculating and explaining energy change

**Working to Grade E**

1 **a** True
**b** 1000 J = 1 kJ
**c** In the equation $Q = mc\Delta T$, $\Delta T$ represents temperature change.
**d** True
**e** In a chemical reaction, energy is released as new bonds are formed.
**f** In a chemical reaction, energy must be supplied to break bonds.
**g** The new pathway has a lower activation energy.

2 **a** F
**b** A
**c** C
**d** Exothermic – energy stored by products is less than energy stored by reactants.

## Working to Grade C

3  **a**  Hydrogen
   **b**  Advantages – water is the only product; hydrogen can be obtained from renewable sources of methane.
   Disadvantages – energy is required to produce hydrogen; few filling stations supply hydrogen gas.
4  14 700 J
5  **a**  2 940 J
   **b**  Exothermic – temperature increased.
6  **a**  **i**  –350 kJ/mol
         **ii**  +400 kJ/mol
   **b**  Reaction 1 – energy stored in reactants is more than energy stored in products.
7

| Reaction | Bonds that break | Bonds that are made |
|---|---|---|
| a | H–H<br>Cl–Cl | H–Cl<br>H–Cl |
| b | H–H<br>H–H<br>O=O | H–O<br>H–O<br>H–O<br>H–O |
| c | C–H<br>Cl–Cl | C–Cl<br>H–Cl |

## Working to Grade A*

8  **a**  O=O
   **b**  Cl–Cl
   **c**  O=O
9  **a**  –185 kJ/mol
   **b**  –483 kJ/mol
   **c**  –115 kJ/mol

## C3 6–8 Examination questions: Calculating and explaining energy change

1  **a**  Keep the cereal the same distance from the water being heated. (2)
   **b**  $\frac{(32+36=38)}{3}$ = 35.3 °C (1)
   **c**  1 mark will be awarded for showing your working, 1 mark for the correct answer.
         $100 \times 4.2 \times 35.3 = 14\,826$ J
   **d**  **i**  During the student's experiment, much energy is transferred to the surroundings, the calorimeter, and the rest of the apparatus. This is not the case for the value on the packet. (1)
         **ii**  Add a lid (or any other suitable suggestion). (1)
2  **a**  **i**  1 mark is awarded for each impact identified, up to a maximum of 2. For example: less methane is released to the atmosphere; less cow manure is available to fertilise crops; less fossil fuel is used; or any other suitable answer.
         **ii**  The farmer will make money from selling the gas. (1)
   **b**  The energy stored by the reactants is greater than the energy stored by the products. (2)
   **c**  Bonds broken:
         4 (C–H) and 2 (O=O) (1)
         Energy required to break these bonds
         $= (4 \times 413) + (2 \times 497)$
         $= 2646$ kJ/mol (1)
         Bonds made:
         2 (C=O) and 4 (H–O)
         Energy given out on forming these bonds
         $= (2 \times 743) + (4 \times 463)$
         $= 3338$ kJ/mol (1)
         Overall energy change for reaction
         = energy required to break bonds – energy given out on forming new bonds
         $= 2646 – 3338$
         $= –692$ kJ/mol (1)

## How Science Works: The periodic table, water, and energy changes

1  **a**  Town D
   **b**  Town A, mean = 5; town B, mean = 28; town C, mean = 7; town D, mean = 58; town E, mean = 7
   **c**  Bar chart, since one of the variables (town) is categoric.
   **d**  Bar chart should include towns A to E labelled on the x-axis; number of drops, with an even scale on the y-axis; correctly plotted bars.
2  **a**  Line graph should include volume of sodium carbonate added ($cm^3$) labelled on the x-axis; number of drops, with an even scale, on the y-axis; correctly plotted data; line of best fit.
   **b**  The value for 7 $cm^3$ of sodium carbonate solution is anomalous.
   **c**  The graph shows that as the volume of sodium carbonate solution added increases, the number of drops of soap solution required for a permanent lather decreases. This indicates that the greater the volume of sodium carbonate solution added, the softer the water becomes. Once 5 $cm^3$ of sodium carbonate solution have been added, the water does not become any softer. This suggests that all the calcium or magnesium ions that were dissolved in the water have now been removed.

## C3 9: Identifying positive ions and carbonates

1  Copper hydroxide – blue; magnesium hydroxide – white; iron (II) hydroxide – green.
2  Dissolve the salt in pure water. Add a few drops of sodium hydroxide solution. If the salt contains aluminium ions, the precipitate will be white. Add excess sodium hydroxide solution to the precipitate. If the precipitate dissolves, the salt contains aluminium ions.

3   Add a few drops of dilute hydrochloric acid to the solid. If it fizzes, test the gas evolved with limewater. If the limewater goes cloudy, the salt was a carbonate.

## C3 10: Identifying halides and sulfates, and doing titrations.

1   Dissolve a little of the solid to be tested in dilute nitric acid. Add silver nitrate solution. If a cream precipitate forms, the salt was a bromide.
2   To improve the accuracy, and so ensure the value calculated is as close to the true value as possible.
3   0.20 mol/dm$^3$

## C3 11: More on titrations

1   0.05 mol
2   0.0048 mol
3   0.104 mol/dm$^3$

## C3 9–11 Levelled questions: Further analysis and quantitative chemistry

### Working to Grade E

1

| Compound of... | Flame colour |
|---|---|
| lithium | crimson |
| sodium | yellow |
| potassium | lilac |
| calcium | red |
| barium | green |

2

| Name of precipitate | Colour |
|---|---|
| aluminium hydroxide | white |
| copper(II) hydroxide | blue |
| iron(II) hydroxide | green |
| calcium hydroxide | white |
| iron(III) hydroxide | brown |
| magnesium hydroxide | white |

3   aluminium hydroxide

### Working to Grade C

4   hydrochloric, barium chloride, white, nitric, silver nitrate, white.
5   Add a few drops of dilute hydrochloric acid to the solid, if it fizzes, test the gas with limewater. If the limewater turns cloudy, the solid is a carbonate.
6   a   Use a **pipette** to measure out exactly 25.00 cm$^3$ of sodium hydroxide solution.

b   Transfer the solution to a **conical flask**.
c   Using a funnel, pour the hydrochloric acid into a burette. Add a few drops of indicator to the **conical** flask.
d   Read the scale on the burette. Add hydrochloric acid from the burette to the sodium hydroxide solution in the **conical flask** until the indicator changes colour.
e   Repeat steps (a) to (d) **until consistent results are obtained.**
7   a   9.70 cm$^3$
    b   9.77 cm$^3$

### Working to Grade A*

8   copper sulfate 4 g/dm$^3$
    sodium chloride 30 g/dm$^3$
    magnesium sulfate 200 g/dm$^3$
9   0.025 mol
10  0.0047 mol
11  58.5 g
12  23.75 g
13  0.924 mol/dm$^3$
14  0.0536 mol/dm$^3$

## C3 9–11 Examination questions: Further analysis and quantitative chemistry

1   a   Dip the end of a clean nichrome wire into the salt. Hold the end of the nichrome wire in a hot Bunsen flame. Observe the flame colour. (1)
    b   3 marks available. 1 will be awarded for each correct alternative conclusion that is suggested and that is supported with evidence from the table. For example:
    The mixture contains calcium ions and lithium ions. Evidence – the flame colour could be caused by both lithium and calcium ions.
    The mixture contains calcium ions and magnesium ions. Evidence – the flame colour could be caused by calcium ions, and the white precipitate could be a mixture of calcium hydroxide and magnesium hydroxide formed when sodium hydroxide solution reacted with calcium and magnesium ions in solution.
    The mixture contains lithium ions and magnesium ions. Evidence – the flame colour could be caused by lithium ions, and the white precipitate could be magnesium hydroxide formed when sodium hydroxide solution reacted with magnesium ions in solution.
    c   Test 4 – sulfuric acid has been added to the mixture, so the test will be positive for sulfate ions, whatever the salt. (1)
    d   Carbonate ions – Test 3, fizzed with acid, and the gas produced made limewater cloudy; (1)
    Chloride ions – Test 5, white precipitate with silver nitrate solution. (1)

**2** Here is a guide to the marking for an essay-style question:

To gain 5/6 marks – All information in the answer should be relevant, clear, organised, and presented in a structured and coherent format. Specialist terms will be used appropriately. Few, if any, errors in grammar, punctuation, and spelling. Answer should include 5 or 6 points from those below.

To gain 3/4 marks – Most of the information should be relevant and presented in a structured and coherent format. Specialist terms will usually be used correctly. There may be occasional errors in grammar, punctuation, and spelling. Answer should include 3 or 4 points of those listed below.

To gain 1/2 marks – Answer will be simplistic. There may be limited use of specialist terms. Errors of grammar, punctuation, and spelling will prevent communication of the science. Answer should include 1 or 2 points of those listed below.

*Points to include:*
- Use a pipette to measure out 25 cm³ of sodium hydroxide solution.
- Transfer the solution to a conical flask.
- Add a few drops of indicator to the solution in the flask.
- Pour hydrochloric acid into a burette and read the scale.
- Allow hydrochloric acid to run into the conical flask, with swirling.
- When the indicator changes colour, the end point has been reached.
- Repeat until several consistent results are obtained, but add acid one drop at a time as the end point is approached.

**3**  **a**  **i**  To find out the approximate volume of acid required. (1)

**ii**  To improve accuracy. (1)

**iii**  $(11.90 + 12.00 + 12.10) \div 3 = 12.00$ (1)

**b**  No – much more DCPIP (80 times) was required with the orange juice than the blackcurrant drink, indicating that the orange juice contained more vitamin C. (1)

**4**  $H_2SO_4 + 2NaOH \rightarrow Na_2SO_4 + H_2O$

Mass of one mole of sulfuric acid
$= (2 \times 1) + 32 + (16 \times 4) = 98\,g$

Concentration of sulfuric acid solution
$= 9.8\,g \div 98\,g = 0.1\,mol$ (1)

Number of moles of sulfuric acid in 23.00 cm³ of solution
$= (23.00 \div 1000) \times 0.1$
$= 0.0023\,moles$ (1)

The equation shows that 1 mole of sulfuric acid reacts with 2 moles of sodium hydroxide. So in 25.00 cm³ sodium hydroxide solution there are 0.0023 × 2 = 0.0046 moles of sodium hydroxide. (1)

So in 1000 cm³ of sodium hydroxide solution there are 0.0046 × (1000 ÷ 25) = 0.184 moles of sodium hydroxide.

So the concentration of sodium hydroxide solution is 0.184 mol/dm³ (1)

## C3 12: Ammonia

**1**  Nitrogen – separated from the air; hydrogen – obtained from natural gas.

**2**  A temperature of 450 °C is chosen for the Haber process. At this temperature, the percentage yield is smaller than it is at lower temperatures. But the rate of the reaction increases as temperature increases. At 450 °C the yield and rate are both acceptable. The environmental impact and energy requirements would be less at lower temperatures, but the rate would be too slow.

**3**  A pressure of 200 atm is chosen for the Haber process. The higher the pressure, the higher the yield. But at high pressures the equipment and operating costs increase. The pressure chosen is a compromise between maximising yield and minimising costs.

## C3 13: Ammonia 2

**1**  An equilibrium reaction is one which can go in both directions. The reactions in each direction happen at the same rate. Equilibrium can only be achieved in a closed system. The reaction below is an example of an equilibrium reaction.
$N_2(g) + 3H_2(g) \rightleftharpoons 2NH_3(g)$

**2**  When an equilibrium reaction is subjected to a change in conditions, the position of the equilibrium reaction shifts so as to counteract the effect of the change. So if the pressure is increased in a gaseous reaction, the reaction that produces fewer molecules is favoured, as shown by the equation for the reaction.

## C3 12–13 Levelled questions: The production of ammonia

### Working to Grade E

**1**

| Raw material | Source |
|---|---|
| nitrogen | separated from the air |
| hydrogen | obtained from natural gas or other sources |

**2**

| | |
|---|---|
| temperature (°C) | 450 |
| pressure (atmospheres) | 200 |
| catalyst | iron |

### Working to Grade C

**3**  1 – A, 2 – B, 3 – C, 4 – E, 5 – D

**4**  **a**  True

**b**  In the reaction vessel, ammonia molecules break down to make nitrogen and hydrogen.

**c**  True

**d**  One mole of nitrogen reacts with three moles of hydrogen to make two moles of ammonia.

**e**  True

## Working to Grade A*

**5**

| Change | Effect on position of equilibrium | | |
|---|---|---|---|
| | Shifts left | No change | Shifts right |
| increasing temperature | | | ✓ |
| decreasing pressure | | | ✓ |
| adding a catalyst | | ✓ | |

**6** Increasing temperature shifts the equilibrium towards the endothermic reaction which absorbs more energy so tending to counteract the effect of increasing the temperature.

Decreasing pressure shifts the equilibrium towards the reaction that produces the higher number of molecules as shown by the symbol equation for the reaction.

There is no effect on the position of the equilibrium when a catalyst is added – catalysts change the activation energy of a reaction, and therefore its rate. They do not affect the position of the equilibrium.

**7**
  **a** True
  **b** True
  **c** At equilibrium, the amounts of the products in the mixture remain the same.
  **d** At equilibrium, the amounts of the reactants in the mixture remain the same.
  **e** True
  **f** True

**8**
  **a** The higher the pressure, the higher the yield of ammonia.
  **b** At this pressure, the yield is acceptable. At higher pressures, the yield would be greater, but the costs of building and maintaining the plant would be much more.

## C3 12–13 Examination questions: The production of ammonia

**1**
  **a**
    **i** The symbol shows that the reaction is reversible. (1)
    **ii** 3, 2 (1)
  **b**
    **i** So that they are not wasted (1) and to reduce costs. (1)
    **ii** To increase the rate of the reaction. (1)
    **iii** 1 mark awarded for each reason, up to a maximum of 2.
      • To reduce the energy requirements of the process – if energy was not transferred from the cooling gases, it would have to be supplied to the reaction vessel from another source.
      • To reduce the temperature of the gases – ammonia condenses at a higher temperature than hydrogen and nitrogen, and so can be separated from the mixture.

  **c**
    **i** As temperature increases, the yield of ammonia decreases. (1)
    **ii** A temperature of 450 °C is chosen for the Haber process. At this temperature, the percentage yield is smaller than it is at lower temperatures. But the rate of the reaction increases as temperature increases. (1) At 450 °C the yield and rate are both acceptable. (1) The environmental impact and energy requirements would be less at lower temperatures, but the rate would be too slow. (1)

**2**
  **a** 1 mark for each of the two true statements identified. True statements: In the equilibrium mixture, sulfur dioxide and oxygen are reacting to make sulfur trioxide.
In the equilibrium mixture, sulfur trioxide is decomposing to make sulfur dioxide and oxygen.
  **b** **Here is a guide to the marking for an essay-style question:**
To gain 5/6 marks – All information in the answer should be relevant, clear, organised, and presented in a structured and coherent format. Specialist terms will be used appropriately. Few, if any, errors in grammar, punctuation, and spelling. Answer should include 5 or 6 points from those below.
To gain 3/4 marks – Most of the information should be relevant and presented in a structured and coherent format. Specialist terms will usually be used correctly. There may be occasional errors in grammar, punctuation, and spelling. Answer should include 3 or 4 points of those listed below.
To gain 1/2 marks – Answer will be simplistic. There may be limited use of specialist terms. Errors of grammar, punctuation, and spelling will prevent communication of the science. Answer should include 1 or 2 points of those listed below.
*Points to include:*
  • If a reaction at equilibrium is subjected to a change, the position of the equilibrium changes so as to tend to counteract the change.
  • If the pressure increases, the equilibrium shifts towards the right.
  • This is because there are fewer molecules shown on the right of the equation.
  • If the pressure decreases, the equilibrium shifts towards the left.
  • If the temperature increases, the equilibrium shifts towards the endothermic reaction (left).
  • This is because the endothermic reaction absorbs the extra energy, so tending to counteract the effect of increasing temperature.

- If the temperature decreases, the equilibrium shifts towards the exothermic reaction (right).
- This is because the exothermic reaction releases energy, so tending to counteract the effect of decreasing the temperature.

## C3 14: Alcohols

1  –O–H

2

| Reacts with... | Products |
| --- | --- |
| oxygen in combustion reactions | carbon dioxide and water |
| sodium | hydrogen and sodium ethoxide |
| oxygen from oxidising agents or by action of microbes | ethanoic acid |

3  $2CH_3OH + 3O_2 \rightarrow 2CO_2 + 4H_2O$

## C3 15: Carboxylic acids

1

| React with... | Products |
| --- | --- |
| oxygen in combustion reactions | carbon dioxide and water |
| carbonates | salt and carbon dioxide and water |
| alcohols | ester and water |

2  To make vinegar; as food additives; as a component of vitamin tablets (ascorbic acid); as a painkiller (aspirin).

3  The pH will be higher for ethanoic acid. pH is a measure of hydrogen ion concentration. The greater the hydrogen ion concentration, the lower the pH. The hydrogen ion concentration is less in ethanoic acid because ethanoic acid does not ionise completely in water. It is a weak acid. Hydrochloric acid does ionise completely in water. It is a strong acid.

## C3 16: Esters

1

ethyl ethanoate

2  ethanol + ethanoic acid → ethyl ethanoate + water

3  Esters are used as solvents in cosmetics and food flavourings.

## C3 14–16 Levelled questions: Alcohols, carboxylic acids, and esters

### Working to Grade E

1

| Compound | Use |
| --- | --- |
| ethanol | alcoholic drinks |
| ethanoic acid | vinegar |
| pentyl pentanoate | flavouring |

2  neutral, hydrogen, acidic, carbon dioxide.

3  Any correct answers are acceptable as examples, including those in the table below.

| Name of group of organic compounds | Examples |
| --- | --- |
| alcohols | ethanol methanol |
| carboxylic acids | ethanoic acid propanoic acid |
| esters | ethyl ethanoate ethyl propanoate |

4  functional, propanol, homologous fuels, solvents, drinks, ethanol, ethanoic, oxidising, microbes

### Working to Grade C

5  a  Carboxylic acids have the functional group –COOH
   b  Carboxylic acids react with alcohols to produce esters.
   c  The molecular formula of methanoic acid is HCOOH.
   d  True
   e  True
   f  Propyl propanoate is an ester.

6

| Name | Molecular formula | Structural formula |
| --- | --- | --- |
| methanol | $CH_3OH$ | |
| ethanol | $CH_3CH_2OH$ | |
| propanol | $CH_3CH_2CH_2OH$ | |
| methanoic acid | HCOOH | |
| ethanoic acid | $CH_3COOH$ | |
| propanoic acid | $CH_3CH_2COOH$ | |

7  a  ethanol, propanoic acid
   b  propanol, ethanoic acid
   c  methanoic acid, methanol
   d  ethanoic acid, ethanol

**8** **a** ethanol + oxygen → carbon dioxide + water
**b** methanol + sodium → sodium methoxide + hydrogen
**c** ethanol + sodium → sodium ethoxide + hydrogen
**d** propanol + oxygen → carbon dioxide + water
**e** ethanoic acid + sodium carbonate → sodium ethanoate + carbon dioxide + water
**f** propanoic acid + calcium carbonate → calcium propanoate + carbon dioxide + water
**g** ethanol + ethanoic acid → ethyl ethanoate + water
**h** propanol + propanoic acid → propyl propanoate + water

## Working to Grade A*

**9** **a** $2CH_3OH + 3O_2 \rightarrow 2CO_2 + 4H_2O$
    **b** $CH_3CH_2OH + 3O_2 \rightarrow 2CO_2 + 3H_2O$
**10** Ethanoic acid does not ionise completely in water. This means it is a weak acid.
**11** **a** Acid A – it has the lowest pH, meaning it is the strongest acid of those in the table.
    **b** Acid C is a weaker acid than acid B, because the pH of acid C is higher.

# C3 14–16 Examination questions: Alcohols, carboxylic acids, and esters

**1** **a**

ethyl ethanoate     (1)

**b** alcohols and carboxylic acids. (1)
**c** To act as a catalyst *or* to speed up the reaction. (1)
**d** It has an interesting flavour. (1)
**2** Uses: fuels, solvents, alcoholic drinks.
**Here is a guide to the marking for an essay-style question:**
To gain 5/6 marks – All information in the answer should be relevant, clear, organised, and presented in a structured and coherent format. Specialist terms will be used appropriately. Few, if any, errors in grammar, punctuation, and spelling. Answer should include 5 or 6 points from those below.
To gain 3/4 marks – Most of the information should be relevant and presented in a structured and coherent format. Specialist terms will usually be used correctly. There may be occasional errors in grammar, punctuation, and spelling. Answer should include 3 or 4 points of those listed below.
To gain 1/2 marks – Answer will be simplistic. There may be limited use of specialist terms. Errors of grammar, punctuation, and spelling will prevent communication of the science. Answer should include 1 or 2 points of those listed below.
*Points to include (for alcoholic drinks):*
- Social disadvantage of drinking alcohol – it slows down reaction times, increasing risks of road accidents.

- Social disadvantage of drinking alcohol – it makes people forgetful, confused, and more likely to act foolishly.
- Social disadvantage of drinking alcohol – it may cause vomiting, unconsciousness, or death.
- Social advantage of drinking alcohol – it makes people feel relaxed for a short time.
- Economic advantages of drinking alcohol – profitable for drinks companies.
- Economic advantage of drinking alcohol – taxes provide income for government.
- Economic disadvantage of drinking alcohol – treating alcohol-related health problems is costly.
- Economic disadvantage of drinking alcohol – dealing with alcohol-related crime is costly.
- Statement evaluating the benefits and costs of drinking alcohol.

**Note:** There are many other possible correct answers to this question.

**3** **a**

(1)

**b** carbon dioxide (1)
**c** Compound A (1)
**d** **i** Only some of its molecules are ionised in solution. (2)
      **ii** $0.1 \, mol/dm^3$ ethanoic acid, (1) since this is a carboxylic acid. In a solution of a carboxylic acid, only some of the ethanoic acid molecules are ionised. (1)

# How Science Works: Analysis, ammonia, and organic chemistry

**1** Questions could include: How many repeats should I do for each acid concentration? What range of acid concentration should I use? What intervals between the acid concentrations would be best?
**2** Independent variable – acid concentration.
Dependent variable – volume of acid required to neutralise $25 \, cm^3$ of $1.0 \, mol/dm^3$ sodium hydroxide solution.
**3** The measurements must measure only the volumes of acid required, and should be as precise as possible.
**4** Suzanna must control all the variables that might affect the dependent variable, including the concentration of sodium hydroxide solution, and the volume of sodium hydroxide solution, and the indicator chosen.
**5** At acid concentration 1.0, the data is clustered closely. At acid concentration 1.5, the data is clustered less closely. This indicates that the data for concentration 1.0 is more precise.
**6** Suzanna might have made mistakes in the investigation; she might have read some values incorrectly; she might have found it difficult to judge the exact level of the acid meniscus on the burette scale; the equipment might not have been clean.

## P3 1: X-rays

1  No need for surgery, so no risk of infection from surgery.

2  Place a sheet of photographic film in black paper. Put the limb between the film and X-ray source. Briefly switch on the source. X-rays will pass through muscle and skin to ionise the film. Film under bone will not be ionised as the bone absorbs X-rays. When the film is developed, the ionised regions turn black, the rest goes white, creating the image.

3  A CT scan can be used to form an image of the tumour. A narrow beam of weak X-rays passes through the patient to be detected by a CCD on the other side. The X-ray source and detector rotate around the body, and a computer analyses the CCD signal to make an image of this slice of the body. To treat the tumour, an intense beam of X-rays is passed through the tumour from many different directions so that the tumour gets a much larger dose than the surrounding tissue – this should kill cells in the tumour without to much damage to healthy cells around it.

## P3 2: Ultrasound

1  A sound wave of frequency above the range of human hearing (20 kHz).

2  Short pulses of ultrasound are fired into the mother towards the baby. As the pulses cross boundaries between tissues (skin, muscle or bone), they are partially reflected back towards their source where they are detected. The time delay between a pulse and its echo can be used to calculate the distance to the tissue boundary which caused the echo. A computer uses the information from the echoes to build up an image of the mother and baby as the ultrasound pulses are scanned across her.

3  133 μs

## P3 1–2 Levelled questions: X-rays and ultrasound

### Working to Grade E

1  atom

2  Ultrasound does not damage the baby's cells by ionisation, but X-rays do.

3  Wear a monitoring badge, have shielding between yourself and the source, keep the X-rays on for as short a time as possible.

4  100 kHz

5  The CT scan uses X-rays passing through a region of the body in many different directions, being detected by a CCD. The film uses X-rays coming from only one direction onto photographic film.

### Working to Grade C

6  Place a sheet of photographic film in black paper. Put the limb between the film and X-ray source. Briefly switch on the source. X-rays will pass through muscle and skin to ionise the film. Film under bone will not be ionised as the bone absorbs X-rays. When the film is developed, the ionised regions turn black, the rest goes white, creating the image.

7  Three from: Wear a monitoring badge. Keep well away because X-rays spread out as they travel away from the source. Have shielding between yourself and the source to absorb the X-rays before they get to you. Keep the X-rays on for as short a time as possible – the longer you are exposed to X-rays, the more chance you have of developing a tumour.

8  A sound wave whose frequency is above 20 kHz, the upper limit of the range of human hearing.

9  A narrow beam of weak X-rays passes through the patient to be detected by a CCD on the other side. The amount of X-rays being absorbed before it reaches the CCD depends on the amount of bone, skin or muscle in the way, so the CCD signal is a measure of the organs between source and detector. The X-ray source and detector rotate around the body, and a computer analyses the CCD signal to make an image of this slice of the body.

10  Short pulses of ultrasound are fired into the mother towards the baby. As the pulses cross boundaries between tissues (skin, muscle or bone), they are partially reflected back towards their source where they are detected. The time delay between a pulse and its echo can be used to calculate the distance to the tissue boundary which caused the echo. A computer uses the information from the echoes to build up an image of the mother and baby as the direction of the ultrasound pulses is scanned across her.

### Working to Grade A*

11  $5 \times 10^{-2}$ m or 5 cm

## P3 1–2 Examination questions: X-rays and ultrasound

1  a  T V R Q S U (4 marks if all correct, 3 marks for one or two mistakes, 2 marks for three mistakes, 1 mark for four mistakes in the order of steps)

   b  For their safety (1). The walls of the room will absorb some of the X-rays (1), causing less damage to her cells by ionisation (1).

2  a  A sound wave (1) with a frequency above 20 kHz (1).

   b  X-rays ionise cells (1), damaging them, sometimes making them into cancer (1). Ultrasound does not ionise cells, so cannot harm the baby (1).

   c  Short pulses of ultrasound are fired into the mother towards the baby (1). As the pulses cross boundaries between tissues, they are partially reflected back towards their source where thay are detected (1). The time delay between a pulse and its echo can be used to calculate the distance to the tissue boundary which caused the echo (1). A computer uses the information from the echoes to build up an image of the mother and baby as the direction of the ultrasound pulses is scanned across her (1).

## P3 3: Refraction

1

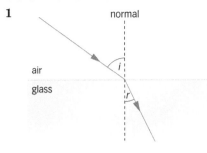

2    The speed of the light changes as it changes medium. In order to keep the frequency the same in both mediums, the wavelength must also change – this usually requires a change of direction.

3    26°

## P3 4: Lenses

1

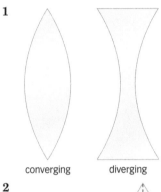

2

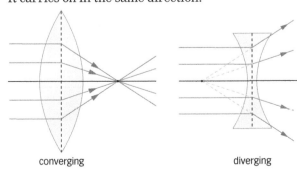

3    Inverted, real and magnified.

## P3 5: Ray diagrams

1    It carries on in the same direction.

2

converging          diverging

3    The image is 2 cm high, real, inverted and 10 cm in front of the lens.

## P3 6: Magnification and power

1    The curvature of its surfaces and the refractive index of its material.

2    +20 D

3    Diverging with a focal length of 10 cm.

## P3 7: Human eyes

1

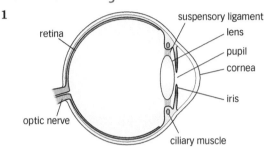

2    The pupil shrinks as the iris contracts to reduce the amount of light reaching the retina, preventing it from being damaged by too much light.

3    People with short-sight have eyes with too high a power. They focus images in front of the retina instead of on it, so a diverging lens placed in front of the cornea can reduce the overall power of their eyes to allow images to be focussed onto the retina.

## P3 8: Cameras

1    The lens refracts light from the object to make a real image on the CCD which records it.

2

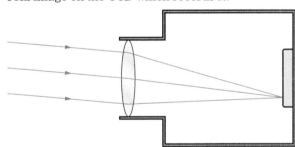

3    The eye lens is made fatter by the ciliary muscles, increasing the power of the lens so that the image is still in focus on the retina. The camera lens is moved away from the CCD to keep the image in focus on it.

## P3 3–8 Levelled questions: The power of lenses

### Working to Grade E

1

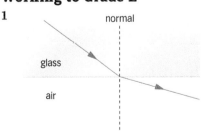

**2**

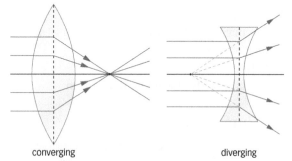

converging                              diverging

**3** The object actually emits light, the image is where the light appears to come from after passing through the lens.

**4** The image is bigger than the object and is upside down.

**5** Real images can be shown on a screen, virtual images cannot.

**6** The image is six times bigger than the object.

**7** The lens can be made thicker to reduce its focal length.

## Working to Grade C

**8** 1.36

**9**

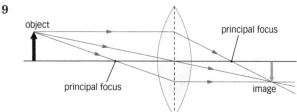

**10**

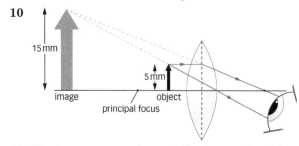

**11** The image moves closer to the lens as the object moves away, so the lens must be moved towards the CCD to keep the image focussed on it.

**12** 10 mm

**13** It protects the retina from damage by too much light. It does this by reducing the size of the pupil as the light intensity increases.

**14**

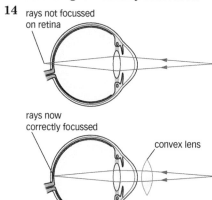

The convex lens increases the power of the eye, allowing the image to be focussed on the retina instead of behind it.

**15** +6.7 D

**16**

suspensory ligament
lens
pupil
cornea
iris
ciliary muscle
retina
optic nerve

## Working to Grade A*

**17** +2.0

**18** The eye lens is made fatter by the ciliary muscles, increasing the power of the lens so that the image is still in focus on the retina. The camera lens is moved away from the CCD to keep the image in focus on it.

## P3 3–8 Examination questions: The power of lenses

**1 a** inverted (1), magnified (1), real (1)

**b i**

magnified
inverted
object
lens
image
real

Straight line (1) between lens and image as shown (1).

**ii** power = $\dfrac{1}{\text{focal length}}$ (1)

focal length = 9.5 squares or 28.5 mm (1)

power = $\dfrac{1}{0.0285}$ = +35 D (1)

**c**

magnified
inverted
object
lens
image
real

Any ray from top of object to lens and then to top of image (1 mark each)

**2** **a** A is iris (1), B is CCD (1) and C is lens (1).
  **b** The shape of the lens can be altered (1).
  **c**

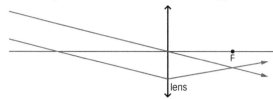

  Rays to the left of the lens parallel (1).
  Rays on the right cross at a focus (1).
  Focus same distance from lens as F (1).
**3** **a** The light slows down (1). Its frequency remains the same (1), so its wavelength must decrease (1), so it has to change direction.
  **b** refractive index $= \dfrac{\sin i}{\sin r}$ (1)
  $= \dfrac{\sin 45}{\sin 30} = \dfrac{0.707}{0.500} = 1.41$ (1)
  **c**

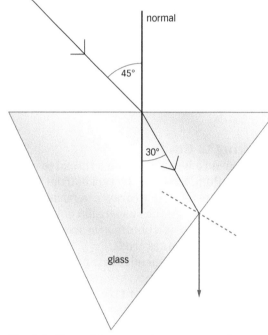

  Straight line with arrow from correct place on glass (1).
  At 45 degrees to surface (by eye) (1).
**4** **a** power $= \dfrac{1}{\text{focal length}}$ (1)
  focal length $= 37.5\,\text{mm}$ (1)
  power $= \dfrac{1}{0.0375} = -27\text{D}$ (1)
  **b**

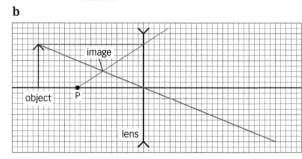

  Ray from top of object through centre of lens (1).

Ray parallel to axis to the lens, then away from P (1).
Correct point labelled image as shown (1).
  **c** magnification $= \dfrac{\text{size of image}}{\text{size of object}}$ (1)
  $= \dfrac{3\ \text{squares}}{8\ \text{squares}} = -0.38$ (1)

## P3 9: Total internal reflection
**1** To look inside people with endoscopes, to carry information in optical fibres.
**2** The light must already be in the slower of the two mediums, hitting the bondary at more than the critical angle.
**3** 46°

## P3 9 Levelled questions: Total internal reflection, optical fibres, and lasers
### Working to Grade E
**1** For optical fibre communications, light shows at concerts.
**2** A thin, long rod of very transparent glass that can carry cable TV signals.

### Working to Grade C
**3**

**4** It doesn't spread out as it travels, so the signal strength doesn't fall rapidly as it travels, allowing a greater distance before the information gets lost.
**5** 1.49
**6** The light must already be in the slower of the two mediums, hitting the bondary at more than the critical angle.
**7** To seal blood vessels by heating them up with pulses of laser light.

### Working to Grade A*
**8** 50°

## P3 9 Examination questions: Total internal reflection, optical fibres, and lasers
**1** **a** refractive index $= \dfrac{1}{\sin c}$ (1) $\sin c = \dfrac{1}{1.6} = 0.625$,
  $c = \sin^{-1} 0.625 = 38.7°$ (1)
  **b**

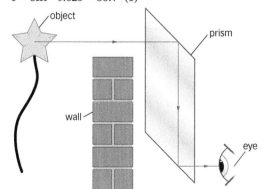

ray continues straight to angled edge of glass (1)
reflects to go vertically down glass (1)
reflects at bottom edge of glass to enter the eye (1)

c   upright (1), virtual (1).

## How science works: Lens magnification

1   The height of the image and distance from the screen from the lens are the dependent variables. The height of the object and the distance from the tracing paper to the lens are the independent variables.
2   The power of the lens.
3   She should calculate the magnification from the heights of the object and image before drawing a scatter-graph of magnification against distance of screen from lens. If she gets a straight line, then her hypothesis is correct.
4   Nina should check the data by doing the experiment again, perhaps by getting a wider range of variables.
5   There will be random errors because she cannot place the equipment exactly as it was last time, zero-errors from measuring from the wrong place (e.g. the edge of the lens instead of the centre), the limited resolution of her ruler (probably 1 mm).

## P3 10: Falling over

1   Halfway down a line drawn from one corner to the opposite corner.
2   Lower the centre of mass, increase the width of the base.
3   No matter how little it tilts, its centre of mass will not be above the surface in contact with the table, so there will be a moment turning it until it hits the table.

## P3 11: Swinging masses

1   The time taken for one complete swing.
2   length
3   0.25 Hz

## P3 12: Moments

1   Anticlockwise
2   8 Nm
3   1.8 m on the right.

## P3 13: Levers

1   Where the lever rests on the triangle of the ground.
2   

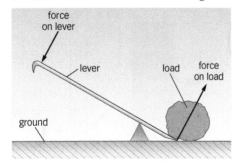

3   For the same applied force, increasing the length of the lever increases the anticlockwise moment on it. When the lever is not moving, the clockwise and anticlockwise moments are the same. An increased clockwise moment means a bigger reaction force from the load if its distance from the pivot does not change.

## P3 14: Hydraulics

1   2000 Pa
2   Air compresses too easily when the pressure increases, whereas oil compresses very little.
3   2400 N

## P3 10–14 Levelled questions: Moments, masses, hydraulics, and levers

### Working to Grade E

1   Halfway down a line drawn from one corner to its opposite across the centre.
2   20 s
3   36 Nm
4   A is clockwise, B, C and D are anticlockwise.
5   Can openers, scissors
6   2000 Pa

### Working to Grade C

7   Suspend a heavy mass on a thread to make a pendulum. Suspend the racket from any point and use the pendulum to draw a vertical line on the racket through the suspension point. Repeat for other suspension points – the centre of mass is where all the lines cross.
8   The mass of the pendulum and the size of its swing.
9   The clockwise moment ($20 \times 1.8 = 36$ Nm) is the same as the anticlockwise moment ($72 \times 0.5 = 36$ Nm).
10  Lower the centre of mass, increase the width of the base.
11  The liquid is incompressible.

### Working to Grade A*

12  60 N
13  1.6 m
14  The nutcrackers are a pair of levers joined at their pivots. You squeeze the long handles together, applying moments to the levers. For each lever to be balanced, the clockwise and anticlockwise moments must be the same, so the moment from your hand is the same size as the moment from the reaction force on the nut. Since the nut is much closer to the pivot than your hand, the force on it must be larger for the moment to be the same.

## P3 10–14 Examination questions: Moments, masses, hydraulics, and levers

**1 a**

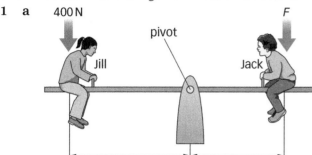

labelling the centre of the see saw (1)

**b** moment = force × distance (1)
= 400 × 2 (1) = 800 (1) Nm (1)

**c** anticlockwise (1), the same as (1)

**d** clockwise moment = anticlockwise moment (1)
$F × 1.5 = 800$ (1)
$F = 800/1.5 = 533$ N (1)

**e** Hang a weight from a string to make a pendulum (1).
Suspend the bat from a point on its edge (1).
Use the pendulum to draw a vertical line through the point of suspension onto the bat (1).
Repeat for other points of suspension, the centre of mass is where the lines cross (1).

## P3 15: Circular motion

1 Towards the centre of the circle upon which that piece of track lies.

2 The centripetal force changes the direction of motion of the object, so that it can follow a circular path.

3 In the absence of any force to the right, you keep on moving forwards as the car turns to the right. As you collide with the left-hand part of the car, it pushes on you, providing the centripetal force which makes you have the same motion as the rest of the car.

## P3 16: Electric motors

1 The field lines are circles in a plane at right angles to the wire, and centred on it.

2 A solenoid only has a magnetic field while there is a current in it. So when the current is turned off, it stops being a magnet.

3 Increase the turrning force on the coil by:
- increasing the current in the coil
- increasing the number of turns of wire in the coil
- increasing the strength of the magnet
- increasing the number of coils.

## P3 17: Transformers

1 Two coils of insulated copper wire wound around a single loop of iron.

2 –0.05 V if the magnet is pulled out at the same speed.

3 230 V

## P3 18: Power supply

1 Reduces the voltage of an a.c. supply and converts it to d.c. supply by using a high frequency transformer.

2 A step-up transformer increases the voltage of a supply, a step-down transformer decreases it.

3 0.33 A

## P3 15–18 Levelled questions: Circular motion and power

### Working to Grade E

1 C

2 Reverse the direction of the current, reverse the direction of the magnetic field.

3 Electric motors, moving-coil meters and loudspeakers.

4 Changes the voltage of an alternating current supply.

5 Step-up.

6 The field lines are circles in a plane at right angles to the wire, and centred on it.

### Working to Grade C

7 **a** increased
  **b** decreased

8 Towards you.

9 To change the direction of the current in the coil every half turn, so that the forces on it always push it round in the same direction.

10 A coil of wire suspended over one end of a bar magnet from the centre of a cone of cardboard which is loosely held in place at its outer edges.

11 5.0 V

12 0.10 A

13 A switch mode transformer is much smaller and lighter, so can be fitted into small power conversion units connected to the mains supply.

### Working to Grade A*

14 An alternating current in a transformer primary coil gives a continually changing magnetic field in the iron core and secondary coil. As the field in the secondary changes, there is a p.d. induced in it. A direct current would not result in any change of magnetic field in the secondary, so there would be no p.d. across it.

15 The p.d. across the secondary coil is less than the p.d. across the primary, yet the power in both coils is the same for an efficient transformer. Since power is p.d. × current, the current in the secondary will be larger than in the primary coil, so the resistance of the secondary needs to be less so that it doesn't get hotter. Increasing the diameter of the wire decreases its resistance.

**16** The brushes allow current to enter and leave the coil as it spins round, via the commutator. The commutator switches the current direction in the coil each time that it does half of a turn, so that the moment acting on the coil always has the same direction.

## P3 15–18 Examination questions: Circular motion and power

**1 a** centripetal (1)

**b** S (1)

**c** friction (1)

**2 a** T R S U Q

all correct (3)

one or two mistakes (2)

three mistakes (1)

**b** $\dfrac{\text{primary turns}}{\text{secondary turns}} = \dfrac{\text{primary voltage}}{\text{secondary voltage}}$ (1)

$\dfrac{\text{primary turns}}{90} = \dfrac{230}{3}$ (1)

primary turns $= \dfrac{230}{3} \times 90 = 6900$ turns (1)

**c** primary current × primary voltage = secondary current × secondary voltage (1)

$I \times 230 = 0.5 \times 3$ (1)

$I \times \dfrac{0.5 \times 3}{230} = 0.0065\,\text{A}$ (1)

**d** It is much more efficient than a normal transformer (1), so waste less energy from the mains supply as heat (1). It is also much lighter (1) and smaller (1) than a normal transformer.

**3** A is commutator (1), B is coil (1), C is magnet (1) and D is brush (1).

## How science works: Electromagnets

**1** The current and the coil turns.

**2** The tightness of the coils, their place on the nail, the diameter and length of the nail, the material of the nail, the weight and shape of each paperclip.

**3** Draw a scatter-graph of the number of paper clips against the current multiplied by the turns of wire and look for a straight line through the origin.

**4** Doing more experiments helps to get around the effects of random errors and the finite resolution of her measurements (she can't get less than 1 paper clip).

**5** It is difficult to wind the wire around the nail exactly the same each time. The ammeter will have a finite resolution, probably to only 0.1 A and different ammeters may give different readings for the same current. She may not be using the same nail or paper clips.

**6** She would need to do an experiment to find out how her results depend on the size of the nail, so that she can extrapolate to a piece of iron the size of the electromagnet she is considering.

# Appendices

## Periodic table

Times of discovery

| before 1800 | 1900–1949 |
| 1800–1849 | 1949–1999 |
| 1849–1899 |

Group

| relative atomic mass | 1.0 |
| atomic number | H |
| name | hydrogen |
| atomic (proton) number | 1 |

| | 1 | 2 | | | | | | | | | | | 3 | 4 | 5 | 6 | 7 | 8 |
|---|---|---|---|---|---|---|---|---|---|---|---|---|---|---|---|---|---|---|
| | | | | | | | | | | | | | | | | | | 4<br>He<br>helium<br>2 |
| Period 2 | 7<br>Li<br>lithium<br>3 | 9<br>Be<br>beryllium<br>4 | | | | | | | | | | | 11<br>B<br>boron<br>5 | 12<br>C<br>carbon<br>6 | 14<br>N<br>nitrogen<br>7 | 16<br>O<br>oxygen<br>8 | 19<br>F<br>fluorine<br>9 | 20<br>Ne<br>neon<br>10 |
| 3 | 23<br>Na<br>sodium<br>11 | 24<br>Mg<br>magnesium<br>12 | | | | | | | | | | | 27<br>Al<br>aluminium<br>13 | 28<br>Si<br>silicon<br>14 | 31<br>P<br>phosphorus<br>15 | 32<br>S<br>sulfur<br>16 | 35.5<br>Cl<br>chlorine<br>17 | 40<br>Ar<br>argon<br>18 |
| 4 | 39<br>K<br>potassium<br>19 | 40<br>Ca<br>calcium<br>20 | 45<br>Sc<br>scandium<br>21 | 48<br>Ti<br>titanium<br>22 | 51<br>V<br>vanadium<br>23 | 52<br>Cr<br>chromium<br>24 | 55<br>Mn<br>manganese<br>25 | 56<br>Fe<br>iron<br>26 | 59<br>Co<br>cobalt<br>27 | 59<br>Ni<br>nickel<br>28 | 63.5<br>Cu<br>copper<br>29 | 65<br>Zn<br>zinc<br>30 | 70<br>Ga<br>gallium<br>31 | 73<br>Ge<br>germanium<br>32 | 75<br>As<br>arsenic<br>33 | 80<br>Se<br>selenium<br>34 | 79<br>Br<br>bromine<br>35 | 84<br>Kr<br>krypton<br>36 |
| 5 | 85<br>Rb<br>rubidium<br>37 | 88<br>Sr<br>strontium<br>38 | 89<br>Y<br>yttrium<br>39 | 91<br>Zr<br>zirconium<br>40 | 93<br>Nb<br>niobium<br>41 | 96<br>Mo<br>molybdenum<br>42 | (98)<br>Tc<br>technetium<br>43 | 101<br>Ru<br>ruthenium<br>44 | 103<br>Rh<br>rhodium<br>45 | 106<br>Pd<br>palladium<br>46 | 108<br>Ag<br>silver<br>47 | 112<br>Cd<br>cadmium<br>48 | 115<br>In<br>indium<br>49 | 119<br>Sn<br>tin<br>50 | 122<br>Sb<br>antimony<br>51 | 128<br>Te<br>tellurium<br>52 | 127<br>I<br>iodine<br>53 | 131<br>Xe<br>xenon<br>54 |
| 6 | 133<br>Cs<br>caesium<br>55 | 137<br>Ba<br>barium<br>56 | 139 *<br>La<br>lanthanum<br>57 | 178.5<br>Hf<br>hafnium<br>72 | 181<br>Ta<br>tantalum<br>73 | 184<br>W<br>tungsten<br>74 | 186<br>Re<br>rhenium<br>75 | 190<br>Os<br>osmium<br>76 | 192<br>Ir<br>iridium<br>77 | 195<br>Pt<br>platinum<br>78 | 197<br>Au<br>gold<br>79 | 201<br>Hg<br>mercury<br>80 | 204<br>Tl<br>thallium<br>81 | 207<br>Pb<br>lead<br>82 | 209<br>Bi<br>bismuth<br>83 | (209)<br>Po<br>polonium<br>84 | 210<br>At<br>astatine<br>85 | 222<br>Rn<br>radon<br>86 |
| 7 | (223)<br>Fr<br>francium<br>87 | (226)<br>Ra<br>radium<br>88 | (227) #<br>Ac<br>actinium<br>89 | (261)<br>Rf<br>rutherfordium<br>104 | (262)<br>Db<br>dubnium<br>105 | (266)<br>Sg<br>seaborgium<br>106 | (264)<br>Bh<br>bohrium<br>107 | (277)<br>Hs<br>hassium<br>108 | (268)<br>Mt<br>meitnerium<br>109 | (271)<br>Ds<br>darmstadtium<br>110 | (272)<br>Rg<br>roentgenium<br>111 | | | | | | |

Elements with atomic numbers 112–116 have been reported but not fully authenticated

*58–71 Lanthanides

| 140<br>Ce<br>cerium<br>58 | 141<br>Pr<br>praseodymium<br>59 | 144<br>Nd<br>neodymium<br>60 | (145)<br>Pm<br>promethium<br>61 | 150<br>Sm<br>samarium<br>62 | 152<br>Eu<br>europium<br>63 | 157<br>Gd<br>gadolinium<br>64 | 159<br>Tb<br>terbium<br>65 | 162.5<br>Dy<br>dysprosium<br>66 | 165<br>Ho<br>holmium<br>67 | 167<br>Er<br>erbium<br>68 | 169<br>Tm<br>thulium<br>69 | 173<br>Yb<br>ytterbium<br>70 | 175<br>Lu<br>lutetium<br>71 |
|---|---|---|---|---|---|---|---|---|---|---|---|---|---|

#90–103 Actinides

| 232<br>Th<br>thorium<br>90 | 231<br>Pa<br>protactinium<br>91 | 238<br>U<br>uranium<br>92 | 237<br>Np<br>neptunium<br>93 | 239<br>Pu<br>plutonium<br>94 | 243<br>Am<br>americium<br>95 | 247<br>Cm<br>curium<br>96 | 247<br>Bk<br>berkelium<br>97 | 252<br>Cf<br>californium<br>98 | (252)<br>Es<br>einsteinium<br>99 | (257)<br>Fm<br>fermium<br>100 | (258)<br>Md<br>mendelevium<br>101 | (259)<br>No<br>nobelium<br>102 | (260)<br>Lr<br>lawrencium<br>103 |
|---|---|---|---|---|---|---|---|---|---|---|---|---|---|

| Equations | |
|---|---|
| $E = m \times c \times \theta$ | $E$ is energy transferred in joules, J<br>$m$ is mass in kilograms, kg<br>$\theta$ is temperature change in degrees Celsius, °C<br>$c$ is specific heat capacity in J/kg°C |
| efficiency $= \dfrac{\text{useful energy out}}{\text{total energy in}}$ ($\times$ 100%) | |
| efficiency $= \dfrac{\text{useful power out}}{\text{total power in}}$ ($\times$ 100%) | |
| $E = P \times t$ | $E$ is energy transferred in kilowatt-hours, kWh<br>$P$ is power in kilowatts, kW<br>$t$ is time in hours, h<br>This equation may also be used when:<br>$E$ is energy transferred in joules, J<br>$P$ is power in watts, W<br>$t$ is time in seconds, s |
| $v = f \times \lambda$ | $v$ is speed in metres per second, m/s<br>$t$ is frequency in hertz, Hz<br>$\lambda$ is wavelength in metres, m |

| Fundamental physical quantities | |
| --- | --- |
| **Physical quantity** | **Unit(s)** |
| length | metre (m) <br> kilometre (km) <br> centimetre (cm) <br> millimetre (mm) |
| mass | kilogram (kg) <br> gram (g) <br> milligram (mg) |
| time | second (s) <br> millisecond (ms) |
| temperature | degree Celsius (°C) <br> kelvin (K) |
| current | ampere (A) <br> milliampere (mA) |
| voltage | volt (V) <br> millivolt (mV) |

| Derived quantities and units | |
| --- | --- |
| **Physical quantity** | **Unit(s)** |
| area | $cm^2$; $m^2$ |
| volume | $cm^3$; $dm^3$; $m^3$; litre (l); millilitre (ml) |
| density | $kg/m^3$; $g/cm^3$ |
| force | newton (N) |
| speed | m/s; km/h |
| energy | joule (J); kilojoule (kJ); megajoule (MJ) |
| power | watt (W); kilowatt (kW); megawatt (MW) |
| frequency | hertz (Hz); kilohertz (kHz) |
| gravitational field strength | N/kg |
| radioactivity | becquerel (Bq) |
| acceleration | $m/s^2$; $km/h^2$ |
| specific heat capacity | J/kg°C |
| specific latent heat | J/kg |

| Electrical symbols | | | | | | | |
| --- | --- | --- | --- | --- | --- | --- | --- |
| junction of conductors | | ammeter | | diode | | capacitor | |
| switch | | voltmeter | | electrolytic capacitor | | relay | |
| primary or secondary cell | | indicator or light source | | LDR | | LED | |
| battery of cells | | or | | thermistor | | NOT gate | |
| power supply | | motor | | AND gate | | OR gate | |
| fuse | | generator | | NOR gate | | NAND gate | |
| fixed resistor | | variable resistor | | | | | |

# Index

Great Clarendon Street, Oxford OX2 6DP

Oxford University Press is a department of the University of Oxford.
It furthers the University's objective of excellence in research,
scholarship, and education by publishing worldwide in

Oxford   New York

Auckland   Cape Town   Dar es Salaam   Hong Kong   Karachi
Kuala Lumpur   Madrid   Melbourne   Mexico City   Nairobi
New Delhi   Shanghai   Taipei   Toronto

With offices in
Argentina   Austria   Brazil   Chile   Czech Republic   France   Greece
Guatemala   Hungary   Italy   Japan   Poland   Portugal   Singapore
South Korea   Switzerland   Thailand   Turkey   Ukraine   Vietnam

Oxford is a registered trade mark of Oxford University Press
in the UK and in certain other countries.

British Library Cataloguing in Publication Data

Data available

ISBN 978-0-19-913594-3

10 9 8 7 6 5 4

Printed in Great Britain by Bell and Bain Ltd, Glasgow

Paper used in the production of this book is a natural, recyclable product
made from wood grown in sustainable forests. The manufacturing process
conforms to the environmental regulations of the country of origin.